Contents

1	**Introduction and background**	1
	1.1. Introduction	1
	1.2. The nature of the polymer molecule	2
	1.3. Molecular orientation	4
	1.4. Crystallinity in polymers	5
	1.5. The glass transition of plastics and rubbers	7
	1.6. Plastics and rubber processing	8
2	**Specimen preparation**	10
	2.1. Introduction	10
	2.2. Sampling	11
	2.3. Microtomy	12
	2.4. Melt pressing and solvent casting	20
	2.5. Preparation of hard and brittle materials	22
	2.6. Preparation for surface studies	24
3	**Surface studies**	26
	3.1. Introduction	26
	3.2. The surface of mouldings and extrudates	26
	3.3. Fracture surfaces	29
	3.4. The surfaces of films	31
	3.5. Pitfalls and precautions	35
4	**Observation of crystalline texture**	37
	4.1. Introduction	37
	4.2. Structure and optical nature of a spherulite	37
	4.3. Characterizing polymer spherulites in the polarizing microscope	39
	4.4. A cautionary note	45
	4.5. Alternative techniques of observation	46
	4.6. Nucleation effects	48
	4.7. Hot-stage microscopy	50
	4.8. An optical diffraction method for the study of spherulites	53

Contents

5 Blends, composites, and copolymers — 55
  5.1. Introduction — 55
  5.2. Blends and random copolymers — 56
  5.3. Soft composites — 56
  5.4. Block copolymers — 60
  5.5. Hard composites — 62
  5.6. Fibre composites — 63
  5.7. Non-fibrous composites — 66

6 Molecular orientation in polymers — 61
  6.1. Introduction — 61
  6.2. Microscopy equipment — 68
  6.3. The aims of birefringence measurements — 68
  6.4. Compensators and compensation — 69
  6.5. The dispersion problem and possible solutions — 72
  6.6. Birefringence measurements on fibres — 73
  6.7. Birefringence measurements on films — 73
  6.8. Birefringence measurements on mouldings — 74

References — 77

Index — 78

Some of the photomicrographs contained in this booklet are reproduced with the kind permission of ICI Petrochemicals and Plastics Division.

1

Introduction and background

1.1. Introduction

A handbook of this size cannot, and should not, attempt to deal with all aspects of the light microscopy of synthetic polymers. It will be clear in what follows that applications for the technique are legion and a prime objective of this text must be to indicate those methods which are most likely to lead to success in relation to specific types of problem. Indeed, the treatment is intended to be a highly practical one which it is hoped will aid those inexperienced in these interesting and diverse materials to utilize their microscope to the greatest effect. Broadly a microscopical examination can be divided into three phases.

First there is the problem of sample selection and preparation. This is dealt with in Chapter 2 and its importance to a successful outcome of the examination is difficult to overemphasize.

The second stage is selection of the microscopical technique. It will be seen that the polymer microscopist may be called upon to use almost any technique of light microscopy. This is of course not to say that every examination requires many techniques. Routine examination of, for example, crystalline texture, may require only a simple transmitted-light polarizing microscope. For the research worker in an industrial or academic laboratory the situation may be very different however, and phase-contrast, dark-field, or interference methods might well contribute more valuable information. For such people the range of available techniques can be seen as an armoury. In practice not all armouries will contain the same weapons but in any event the users are still faced with the challenge of choosing the best weapon for the battle in hand. If this seems a somewhat militaristic approach it perhaps reflects the stout opposition to examination put up by many polymer specimens!

Interpretation of the image represents the third stage of our examination. This can be considered in two steps. First, in order to understand what the image is saying about the chosen sample, we must understand the image-formation characteristics of the microscope we are using and the relevant light—matter interactions taking place in the specimen. This text does not explore these in any detail except insofar as the interpretation is in some way special to polymers. Readers are referred to other handbooks in this series for details of the microscopical methods used and to books and papers listed in the references. The second part of interpretation is to relate what is seen in the microscope to the specimen as a whole and to the problem being investigated. In case this step is thought to be divorced from true image interpretation it should be born in mind that understanding the nature and relevance of what is being seen may well suggest further, and perhaps more revealing, methods of examination or sample preparation. It is for this reason that the polymer microscopist, particularly in the industrial context, is well advised to have available a certain amount of background information on the specimens being examined. This will result in the information extracted from images being especially relevant.

This handbook covers the basics of specimen preparation, the microstructure of semicrystalline polymers, blends, and composites, and methods for the study of polymer surfaces. The coverage is generally qualitative rather than quantitative, except for the chapter on molecular-orientation effects, where birefringence measurement is treated in some detail.

Important and practically significant areas of polymer microscopy which are felt to be outside the scope of this text, either on account of their heavy demands on theory or because of the specificity of their field of application, include measurements using transmitted-light interferometry, u.v. microscopy, photometry, and fluorescence microscopy. These are the province of a larger text.

References to material in the published literature have been avoided but a short list of references is included.

1.2. The nature of the polymer molecule

By definition a polymer is a large molecule (a macromolecule) composed of a repeating sequence of smaller and simpler chemical units. The molecular weight is of the order of many tens of thousands and the molecules, in the form of a long chain, may be linear, branched, or cross-linked. Such a description will be familiar to the biologists but in this text we are concerned not with biological materials but with the synthetic macromolecules more popularly known as 'plastics' and 'rubbers'. The distinction between these two kinds of synthetic polymer, which is made mainly on the basis of their physical properties, is artificial and becoming increasingly blurred. It is easier therefore to use the more general term *polymers* and in the present context to allow ourselves the convenience of dropping the term 'synthetic'.

Variations in molecular structure, configuration, mass, and arrangement dictate the physical properties of a polymer. In turn these control the specimen preparation techniques applicable and, to a certain extent, the microscopical examination techniques that can be employed. In view of the fundamental similarity with 'natural' polymers it is not surprising that many of the methods of preparation (but not all) are similar to those used by the biologists.

We can readily distinguish two classes of polymer. On one hand we have the *thermoplastics*. When heated these soften, become rubbery, and eventually flow as a viscous and rather elastic fluid. The ease of flow will vary with molecular mass, chemical composition, and the extent to which the molecules are branched. On cooling they can be returned to their original form. In principle this cycle of heating and cooling can be repeated indefinitely and the material always returned to its original state. In practice *degradation* of the material may occur — especially rapidly if overheated. Those properties of the material, such as strength, which depend on high molecular weight, are modified, often adversely. This may have consequences in terms of the specimen-preparation method used for microscopy. A physical property *not* affected by modest degradation is the refractive index of the thermoplastic. Thus degraded material is not identifiable by refractive-index assessment in the microscope unless it is particularly severe; by which time its condition will usually be indicated by other characteristics such as darkening in colour. Typical thermoplastics include polyethylene, nylon, poly(vinyl chloride), polypropylene, and polystyrene.

The other class of polymers is the *thermosetting* materials. Thermosets flow on heating, but then harden, and flow ceases. This occurs as a result of chemical cross-linking and is irreversible. Subsequent cooling and re-heating has no apparent effect on the nature of the material unless temperatures are reached which destroy the chemical structure. Typical thermosetting polymers such as phenol formaldehyde or melamine formaldehyde resin are hard, brittle materials which microscopically are treated more like metals or ceramics than biological materials.

Molecular cross-linking also governs the properties of certain rubbers, including natural rubber. Once 'cured' or *vulcanized*, the properties of the material are permanently modified.

To a certain extent polymer molecules can be 'tailored' by the chemists to meet the needs of specific applications. On the other hand commercial considerations militate against the production of a vast range of different polymers. However, modification of a polymer is possible by changing the molecular-weight distribution, blending with other polymers, or 'reinforcing' with inorganic materials often in the form of fibres. In practice therefore, commercial polymers may contain several phases. These phases will often be of a size accessible to the light microscope.

Another method of modifying a polymer is to produce a *copolymer*. A molecule consisting of only one chemical type of repeat unit is called a *homopolymer*. A homopolymer is produced from a single starting organic chemical compound called a *monomer*. Polyethylene, polypropylene, and PVC are examples of homopolymers. A copolymer has units of a different chemical structure inserted into its molecules. This insertion may be random, or regular so that *blocks* or runs of similar chemical structure occur. The number and sequence of introduced units will effect the physical properties. It is often possible to investigate block copolymers with a light microscope since the molecular blocks can come together to produce *domains* of a size within the resolving range. In certain cases however, these are better investigated using transmission electron microscopy (TEM). Random copolymers will normally have a different refractive index from their homopolymer counterparts and may be identifiable in the microscope on this basis; they do not however show any phase structure.

In use some plastics and most rubbers would not survive for very long without additives intended to *stabilize* their properties against the environment, particularly u.v. radiation. Other stabilizers may be necessary to protect the materials from thermal decomposition during the production of fabricated products. Pigments

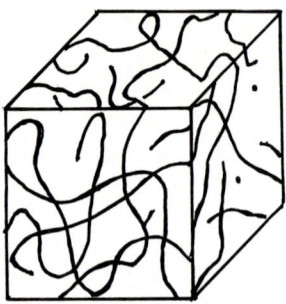

Fig. 1.1. Arrangement of polymer molecules in the melt or in an amorphous solid.

are added to give an almost limitless range of colours and *fillers* may be introduced to reduce material costs. In practice the microscopy of polymers is therefore not the microscopy of pure polymers but of quite complicated assemblages of additives more or less well dispersed in the polymer itself. The study of the distribution and dispersion of these additives is fertile ground for the light microscopist.

1.3. Molecular orientation

The organization of the long-chain molecules in the bulk material has a profound effect on the physical, including optical, properties. In dilute solution in solvent, polymer molecules (which may be up to the order of 10 μm long) occur as randomly coiled flexible chains, the volume occupied by each chain depending on the solvent–polymer interaction. In the molten state, polymer chains are again flexible, in motion but entangled. The somewhat repulsive concept of a can of live worms has been used to aid visualization of this situation, although it has been pointed out that the length-to-thickness ratio would be about 10 m to 5 mm! A similar random organization of molecules may exist in the solid state and is represented in Fig. 1.1. The optical properties of such an arrangement will be independent of position in the material or the direction of observation. The material is characterized by a single unique refractive index and may be described as being both *homogeneous* and *isotropic*. Note that in a bulk specimen consisting of many entangled and randomly arranged molecules, this is true regardless of any difference there might be in the optical characteristics of individual molecules between directions along and across their length. Indeed most polymer molecules do exhibit such a difference and are said to show anisotropy of polarizability.

Light microscopy applied to a specimen of such an idealized homogeneous isotropic specimen would yield no useful information as no contrast is available for image production. As pointed out above, however, most 'real' specimens would at least contain additives which might interest the microscopist.

Consider now a situation in which the organization of molecules is nonrandom. Fig. 1.2. shows a possible arrangement in which there is a preferred direction of

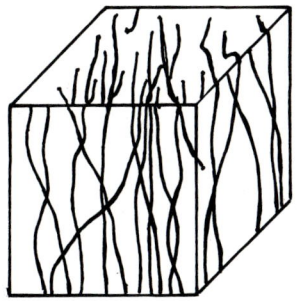

Fig. 1.2. Oriented polymer molecules.

molecular alignment. The alignment is by no means complete but, assuming that the molecules themselves show anisotropy of polarizability, the properties of this region of the specimen depicted in Fig. 1.2 will depend on the direction of measurement or observation. In particular the refractive index of the material will be dependent on the path of light passing through the material. Such a material is termed *doubly refracting* or, more loosely, *birefringent*. Specimens of this type are well known to mineralogists, petrologists, and others. The double-refraction effect, or optical anisotropy, demonstrated by this imaginary specimen permits the generation of contrast in a polarizing microscope. Note that the specimen is strictly still homogeneous since it consists of only one phase. Nevertheless, a means of contrast generation has been produced. The magnitude of the optical anisotropy will depend on the degree to which the molecules are aligned. Randomness of arrangement and total alignment represent extremes of a scale; at one end the specimen will show no anisotropy (i.e. it is isotropic), at the other a maximum. Observations and measurements using a polarizing microscope can thus provide information on molecular arrangement. This is discussed further in Chapter 7, and is of considerable practical significance since many processes impart considerable molecular orientation; in some cases by default, in others by design.

1.4. Crystallinity in polymers

Many polymers are partially crystalline but only in a few exceptional cases does the degree of crystallinity approach 100 per cent. More commonly about 50–70 per cent of the material is crystalline and the most direct evidence for this comes from X-ray diffraction experiments. Sharp diffraction peaks in the X-ray pattern show regions of three-dimensional order whilst additional diffuse rings are evidence for some of the polymer being in a liquid-like state.

Semicrystalline polymers are those whose molecules show regularity both in chemical structure and in the geometrical arrangement of the chemical groups present. Any regularity of side-groups attached to the molecular chain is fixed at the polymerization stage in the manufacture of the polymer. Those molecules having a random arrangement of side-groups are termed *atactic;* those with a given side-group always on the same side of the chain are termed *isotactic,* and, if the position of a side-group regularly alternates, the molecules is said to be *syndiotactic.* Note that a given polymer may exist in more than one structural arrangement. Thus polypropylene may be produced as isotactic or atactic material. Because of the structural regularity, the isotactic material crystallizes readily and isotactic polypropylene is one of the more interesting polymers which can be examined in the polarizing microscope. On the other hand, atactic polypropylene cannot be induced to crystallize and is of less interest. The differences in physical properties produced by crystallization are profound and in practice all polypropylene products, mouldings, films, and fibres, are based on highly isotactic polypropylene. Atactic polypropylene by contrast is a rather sticky rubbery material of little commercial significance.

Other factors besides tacticity control the degree of crystallinity which can be achieved in practice but tacticity predominates over such factors as polarity and chain stiffness.

Polymer single crystals large enough to be resolved in the light microscope can be grown from dilute solutions (e.g. polyethylene, poly (4-methyl pentene-1)) but crystallization from the melt does not normally yield individual crystals on this scale. Under these less favourable growth conditions, the regions of crystallographic regularity form radially symmetric polycrystalline arrays called spherulites. These may be many micrometres, even millimetres, in diameter. These are discussed in Chapter 4 and it is sufficient here to indicate that spherulites are a major microstructural element as far as the light microscopist is concerned and their study is rewarding both in understanding the consequences of different manufacturing processes and in predicting the behaviour of manufactured products.

The arrangement of the molecules within a crystalline region of a polymer was over past years a matter of considerable controversy. The broad diffraction peaks observed in X-ray patterns could be explained by either a small crystallite size or the presence of lattice defects. Unfortunately the patterns were too weak to allow distinction between these possibilities. The early fringed-micelle model was chosen to explain the X-ray observations. This envisaged structure is shown in Fig. 1.3. A single molecule is seen to take part in the generation of several ordered regions, between which are the non-crystalline or amorphous parts of the material. The micelle regions were considered too small for microscopic observation.

The production of polymer single crystals from solution and the evidence that the same form of lamella crystals existed in polymer bulk crystallized from the molten state changed the model substantially. X-ray and electron-diffraction evidence showed that although the molecules were almost perpendicular to the plane of the tabular single crystals, the thickness of these crystals was much less than the molecular length. Chain folding and re-entry of the molecule into the same crystal was proposed to explain these observations and supported by much experimental evidence. This revised model shown in Fig. 1.4 is now accepted as an alternative mode of crystallization in bulk plastics crystallized from the melt. These

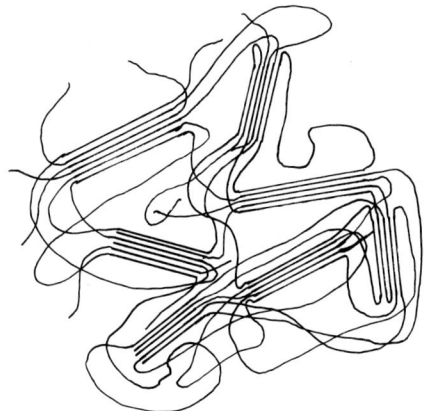

Fig. 1.3. The fringed-micelle model.

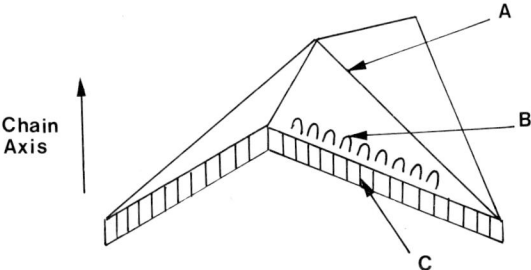

Fig. 1.4. The folded chain crystal model. (A) The ridge normally collapses during specimen preparation for microscopy leading to a 'pleat' across the crystal; (B) chain folds in the crystal surface; (C) orientation of polymer chains.

crystals or lamellae grow in the radially organized manner to produce the spherulites referred to above. The single crystals suffer many of the types of crystallographic defect recognized in non-polymeric materials. To these may be added defects caused by chain ends, folds, and tacticity irregularities.

Other modes of organization can occur, such as extended chain crystallization. These manifest themselves under 'abnormal' conditions of crystallization pressure or temperature and in the presence of molecular orientation. Some knowledge of these basic concepts of polymer crystallization is essential to the light microscopist, even though much is on a scale below the limit of resolution of the instrument. The observation of spherulites and allied structures form a substantial part of polymer microscopy and this basic knowledge aids the interpretation of what is seen.

1.5. The glass transition of plastics and rubbers

If a non-crystalline (or *amorphous*) thermoplastic polymer is cooled, a temperature is reached at which the material takes on the stiffness, hardness, and, particularly, the brittleness of a glass. The precise temperature at which this happens is ill-defined and to some extent depends on the method of measurement. The significance of this glass transition temperature (T_g) to the microscopist is mostly in terms of specimen preparation (see Chapter 2). It might usefully be added that the rate of change of refractive index with temperature (dn/dt) itself changes markedly at T_g.

Semi-crystalline materials, because they contain a proportion of amorphous material also exhibit a T_g. For these materials T_g is virtually independent of the degree of crystallinity, but as the amount of amorphous material may be quite small (< 20 per cent) the practical significance of T_g is reduced.

1.6. Plastics and rubber processing

The microstructure of plastics and rubbers is greatly influenced by manufacturing processes. This is one of the reasons why light microscopy of these materials is particularly rewarding.

The main processes, which normally start from granules or powders, are
1. Extrusion;
2. Injection moulding;
3. Vacuum forming;
4. Blow moulding;
5. Compression moulding;
6. Calendering.

These processes may be preceded by mixing operations such as dry blending or milling to introduce additives, and followed by secondary processes such as coating, painting, heat sealing, welding, lamination, or trimming. Vacuum forming is preceded by the production of sheet.

A minority of products are produced by casting a liquid which polymerizes *in situ* within a mould with the aid of a catalyst.

Certain products have their own specialized areas of technology. These include the production of fibres, films, and foams.

Such a diversity of processes leads to a diversity of microstructure within the reach of light microscopy. All the primary processes, with the exception of compression moulding, are capable of producing high degrees of molecular orientation. Extrusion, injection moulding, blow moulding, and calendering may in addition modify the distribution and dispersion of additives or phases in the polymer.

Most of the primary processes take the material well above its melting point and the cooling rate, induced orientation, and applied pressure will modify the crystalline morphology, the degree of crystallinity, and its distribution in the final product.

Secondary processes such as welding or heat sealing will re-melt part of the product, then allow recrystallization to a different degree of crystallinity and a different morphology. Solvents in paint or lacquers can affect surfaces. Morphology may be modified by cold drawing or the effect of aggressive environments. The finished plastics or rubber product will normally have endured a series of operations, each of which affects the microstructure. It is therefore not surprising that this can show considerable complexity.

2

Specimen preparation

2.1. Introduction

Good specimen preparation is essential if the best possible results are to be obtained from the microscope. At the very best, morphological information will be lost if the preparative techniques are inadequate. At worst, artefacts will mislead and produce a completely wrong interpretation of what the specimen has to show.

There is very little published literature on the details of specimen preparation of polymers. This is partly because many polymers are comparative newcomers to the field of materials microscopy but also because differences in crystallinity, additive content, and the physical form of specimens make specific 'rules' for preparation difficult if not impossible, to formulate. The only exception to this is in the field of fibre microscopy where some well-tried methods are firmly established. This chapter is a guide to some of the basic methods of specimen preparation. It will almost always be necessary to modify the detail of the methods to deal with the particular material in hand.

Plastics exhibit a wide range of chemical and physical properties. They may be hard or soft, brittle or tough, so that it is not surprising that it is necessary to call upon a variety of techniques in preparing specimens for transmitted or reflected light microscopy. The three main factors dictating the choice of technique are

1. The physical and chemical properties of the polymer (e.g. hardness, stiffness, solubility);
2. The aspect of the polymer texture being studied (e.g. crystalline texture, phase separation, additive content);
3. The form of specimen (e.g. granules, fibres, film).

Fig. 2.1 illustrates the range of mechanical properties of some synthetic polymers. The polymers included are intended only to be a representative sample and the properties, hardness and toughness, are not rigorously defined. For example, *toughness* is defined in terms of either breaking stress or yield stress according to the normal mode of failure of the material when strained fairly rapidly at room temperature. Nevertheless, the diagram can be divided into zones within which a particular preparation technique predominates.

In the upper right region of Fig. 2.1. are those materials which are most conveniently prepared for reflected-light examination using techniques similar to those used for metals. If transmitted-light work is necessary, petrological methods are normally used to obtain thin sections.

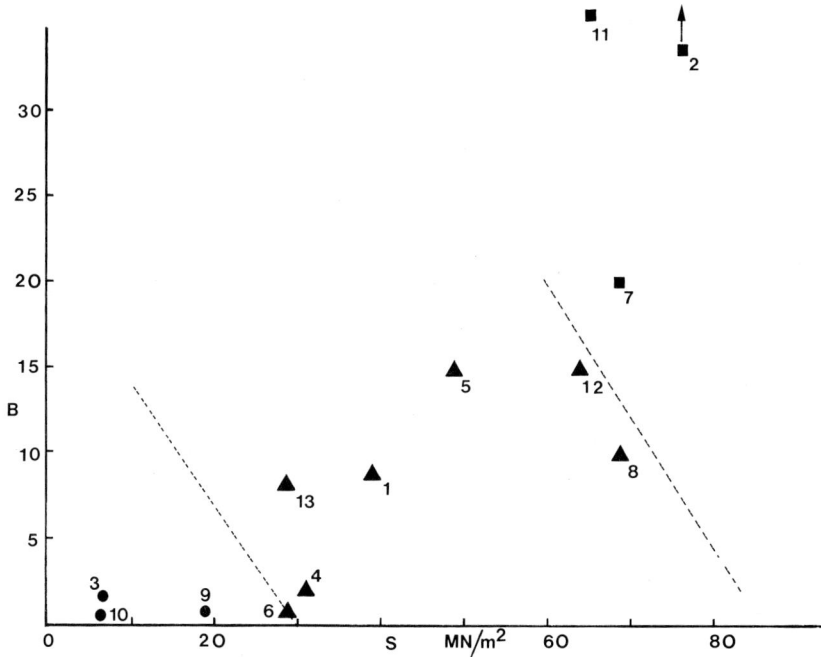

Fig. 2.1. Relative hardness (B) and Strength (S) for a selection of polymers. (▲) Sections easily; (●) specimen needs cooling; (■) specimen needs heating or an alternative technique. 1. Polystyrene; 2. polyimide; 3. L.D. polyethylene; 4. H.D. polyethylene; 5. rigid poly(vinyl chloride); 6. plasticized poly(vinyl chloride); 7. poly (methyl methacrylate); 8. nylon 6; 9. butyl rubber; 10. nitrile rubber; 11. phenol formaldehyde resin; 12. poly acetal; 13. polypropylene.

The central region of Fig. 2.1. contains those materials which can be thin-sectioned by microtomy at, or about, room temperature. The lower left zone contains the 'problem' polymers which are so soft and deformable that they require special low-temperature techniques for microtomy.

It should be stressed that a polymer which normally lends itself to microtomy will not necessarily be sectionable by that technique if highly filled with inorganic additives such as glass fibre, talc, or calcium carbonate. However, normal levels of soft pigments, u.v. stabilizers, lubricants, etc. do not significantly modify the sectioning behaviour.

2.2. Sampling

A common feature of all specimen preparation is the initial selection of the specimen. The statistical significance of even routine tasks in polymer microscopy deserves careful consideration.

In an experiment to determine the particle-size distribution in polymer latex, one might measure 5000 particles. In the original manufacturing autoclave there might well have been 10^{10} particles. Clearly, positive efforts must be made to ensure that those measured are typical of the whole. Sampling will probably be done in a series of stages; for example,

$$\text{autoclave} \rightarrow \text{drum} \rightarrow \text{1 litre bottle} \rightarrow \text{pipette.}$$

At each stage attempts to overcome sedimentation effects must be taken. Unfortunately, it is often the case that not all stages of sampling are under the control of the microscopist.

Similar, but not quite so severe, problems exist when sampling polymer powders, such as PVC, or particulate additives. *Cone and quartering* is the simplest method of sampling. The entire mass of powder to be sampled is poured on to a clean surface so as to produce a cone. This is then divided into four by vertical 'cuts' through its centre. Opposite quarters are then recombined and the whole process repeated until the required size of sample has been obtained. Alternative methods of sampling powders involve riffling equipment. The basic principle is that a supply of the powder to be sampled is fed down a chute into a series of containers which pass continuously beneath it. Usually these containers are held in a rotating disc so that each passes regularly under the chute. To subdivide the powder further, the experiment is repeated with the contents of a single container.

Sampling large solid objects for microtomy or other preparative methods can also pose problems. Being asked to assess the pigment distribution in the wall of a full-sized plastic dustbin is a nonsensical request. It is obviously impracticable to section the entire bin! Further information on the need for the assessment might locate an area of the bin of particular interest. If no further information is available, then several randomly chosen areas should be examined or, with some experience, it might be possible to guess, by taking into account how the bin has been made, the likely whereabouts of the best and worst pigment distributions.

An indication of the method of sampling should be included in any report on the microscopy of such specimens.

2.3. Microtomy

The aim is to produce thin slices of material between 1 and 40 μm thick for transmitted light observation. For routine observations, 5 μm is usually satisfactory, and is also reasonably easily achieved with a standard microtome, usually designed for use on biological materials. For high-resolution microscopy and for certain quantitative observations, thinner sections are desirable. Given any choice in the matter the area of the section should be kept large, say up to 10 × 10 mm, but often the form of the specimen or difficulties in cutting limit this to only a few square millimetres. Obviously the smaller the area of a section, the lower the statistical significance of any conclusions drawn from it.

The suitability of a section for subsequent microscopy is judged not only on the basis of thickness but also on quality. Sections should be strain-free and free from marks resulting from knife-edge imperfections (longitudinal marks) or judder (transverse marks).

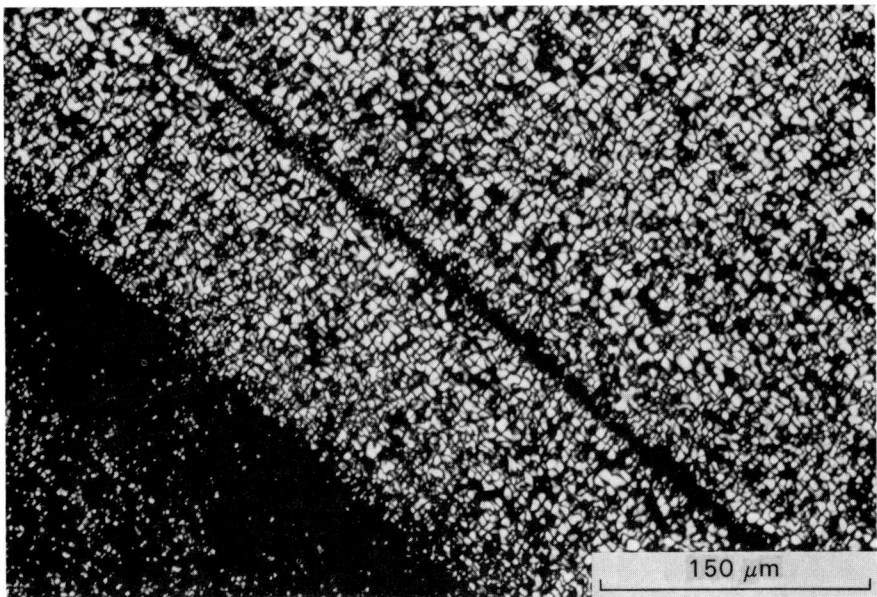

Fig. 2.2. Good section of polypropylene (crossed polars).

Excessive residual strain renders a section unsuitable for polarized-light or interference microscopy and creates difficulties in handling, as differential strain will encourage the section to roll up or buckle. For this reason satisfactory photomicrography of strained sections is almost impossible. Figs. 2.2 and 2.3 shown two consecutive (*serial*) sections of part of a polypropylene moulding, one more carefully cut than the other. The poor section illustrates all the defects mentioned above and is virtually useless for subsequent critical examination.

Techniques have been described in the literature for 'relaxing' residual *cutting strain* in plastics and rubber sections. These techniques rely on either heat or chemical attack to allow the relaxation to take place. Samples of polyethylene may, for example, be removed from the microtome knife and floated on the surface of hot glycerol, and many rubbers can be relaxed by smoothing the section with a paintbrush soaked in xylene. Generally such procedures are to be avoided for the following reasons.

1. Differences in molecular orientation may exist in the specimen and may be relevant to the microscopical examination. The relaxation technique can destroy valuable evidence since the orientation can be annealed away or at least reduced. Note that if sections tend to curl or buckle due to gradients of molecular orientation, this effect is reduced if the sections are cut *thinner*.

2. Thermal methods may produce crystallographic changes. Furthermore, in multiphase materials, such as some propylene–ethylene copolymers, the distribution of the lower melting point phase can be inadvertently modified.

14   *Light microscropy of synthetic polymers*

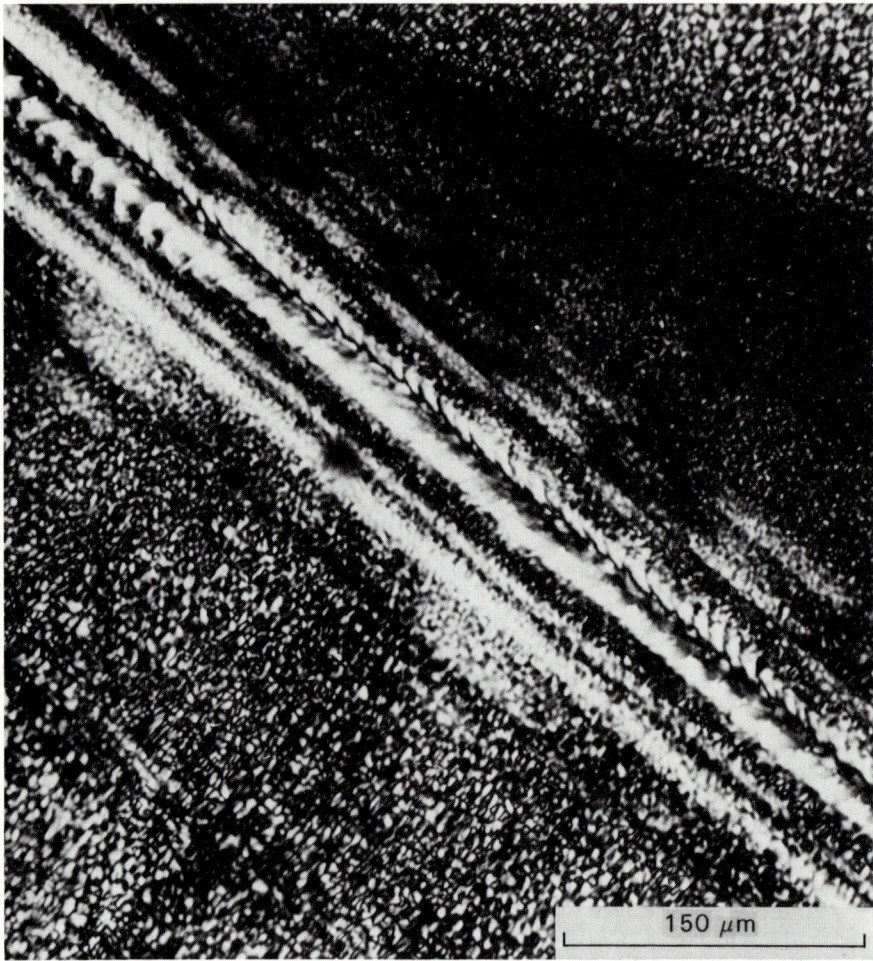

Fig. 2.3. Poor section showing cutting faults (crossed polars).

3. Chemical methods usually involve swelling of the section by an uptake of solvent or plasticizer. This can modify the arrangement of phases in an uncontrolled way and also might give rise to problems when the section is subsequently mounted (see below).

4. Many plastics and rubbers are particularly open to chemical attack as thin sections. Manufacturers data proclaiming a liquid safe for use in contact with the polymer may not apply when dealing with a 1 μm thick section.

5. Strained sections are usually the result of a blunt microtome knife. Prevention is better than cure.

Finally, it should be born in mind that polymers generally have a high coefficient of expansion. Thus a 10 mm thick specimen will expand about 2 μm for every 1°C temperature rise. Reasonable control of the temperature of the environment containing the microtome is clearly desirable.

Choice of microtome and knife

Extreme rigidity is required for sectioning plastics and microtomes of the *base sledge* type are recommended for tough materials or for large area sectioning. It is essential that the knife can be firmly clamped and that the sledge is free running. Rotary types of microtome are also suitable for the less difficult specimens.

A range of steel knives of different profiles is available from manufacturers. In most instances the *plano-plano* ('C') and *chisel-edge* ('D') profiles yield the best results, particularly with harder materials. Plano-concave knives are generally unsatisfactory, having an insufficiently rigid leading edge unless cutting fibres or sections of thin films or foils. The most common reason for difficulty in obtaining sections is a blunt knife.

Extreme knife sharpness is required for all polymer microtomy. Various machines can be used to give a satisfactory knife edge, but a final manual polish using a glass plate and a fine abrasive, such as 0.25 micron diamond paste is recommended. Because of the ease of deformation of most polymers and because this becomes very evident in a polarizing microscope, specimens to be examined by this technique require special care. The standard of knife edge is especially critical and must be as free from imperfections as possible. Glass knives are satisfactory for most sectionable polymers provided the specimen does not exceed about 2 mm in width. Freshly made glass knives with an angle around 45° are almost essential for thin sections of 2 μm thickness or less.

Conversely, tungsten carbide-tipped knives are particularly satisfactory for cutting hard materials or large areas of the tougher polymers. The cutting edge remains reasonably usable for much longer than a standard steel knife but the sharpness, as distinct from the straightness of the edge, is inferior to what can be obtained by lapping a standard knife edge just prior to use.

The accurate setting of the optimum knife inclination is important and the specimen must be advanced a few micrometres at a time, especially when the first sections are being cut. No general rules for setting the knife inclination can be given except that harder materials require steeper inclinations. It is always advisable to experiment with different angles to obtain the optimum.

Holding the specimen

The specimen may be clamped directly but avoiding excessive deformation, into the microtome jaws. Very thin or small specimens may be supported by prior embedding in acrylic or epoxy resins. Such techniques may be used for fibres, films, and powder particles. Care should be taken not to overheat the specimen if the embedding material requires heating or is highly exothermic during the curing reaction. Film samples may be gripped between two sheets of polythene about 3 mm thick.

Impregnation, aided by a cyclic variation of pressure, may ease the microtomy of specimens such as foams having a cellular or 'open' structure. It should always be checked that any impregnation or embedding medium does not interact with the polymer being studied. Holding the specimen by freezing it in ice offers the advantages of speed and freedom from specimen damage, but introduces the need to dry the section before mounting.

Low- and high-temperature sectioning

An ideal homogeneous isotropic amorphous polymer is best sectioned a little below its T_g. In practice various factors modify the situation. These include

1. Crystallinity and/or cross-linking;
2. Additive content;
3. Thickness of section required and area of section;
4. Type of knife and knife profile.

The effect of the degree of crystallinity is well illustrated by polyethylene. Low-density polyethylene (LDP) is difficult to section at room temperature without introducing considerable deformation. This introduced strain manifests itself clearly when the specimen is examined between crossed polars and can render the section useless. To microtome LDP easily it is necessary to reduce its temperature to around the T_g of the amorphous region of the material ($-120°C$). High-density polyethylene (HDP) has a higher degree of crystallinity and is thus harder at room temperature. Although the T_g of the amorphous regions is much the same as for LDP, HDP is considerably easier to section at room temperature. In practice both materials benefit from some measure of cooling. It is often inconvenient to go as low as the T_g but some improvement in section quality is noticeable at temperatures below about $-50°C$; a temperature which is much more readily obtained and maintained using commercially available equipment.

The inclusion of additives in materials may improve or mar the cutting performance of the material. Lower cutting temperatures will, for example, be necessitated by the inclusion of plasticizer in poly vinyl chloride (PVC). The effect of the plasticizer is in effect to drive T_g down from about $+80°C$ to below room temperature. On the other hand, particulate additives, particularly if small and well distributed, may harden the polymer and improve the cutting characteristics.

A moderately 'difficult' polymer may allow sectioning at room temperature at a thickness of, say, 10 μm but thinner sections may need cooling. The shape and size of the specimen also becomes important in this situation and thinner sections will usually be more readily obtained if the area cut is reduced to the minimum acceptable on other counts.

The practical problems of cooling specimens and possibly the microtome down to about $-50°C$ are not severe and are solved by most microtome manufacturers. For modestly low temperatures the whole environment of the specimen and knife is cooled by inclusion of these in a cold compartment. However, very satisfactory results can be obtained using the standard type of specimen stage cooled by the expansion of liquid CO_2. Temperatures as low as -60 to $-80°C$ can be achieved by these means and are suitable for at least improving the sectioning of most difficult materials (with the notable exception of silicone rubbers). In many cases, including the sectioning of butyl and natural rubbers, cooling the knife is not necessary, at least for routine work.

Cooling to lower temperatures still normally requires the use of liquid nitrogen. Various rather primitive systems of cooling specimens (and sometimes knives) with liquid nitrogen have been developed in individual research laboratories, and by some microtome manufacturers. Temperatures around $-160°C$ have been achieved and the benefits for working with such materials as LDP, silicone rubber, and highly plasticized PVC have been well established. It should, of course, be added that sectioning at these temperatures is well catered for by manufacturers of ultra-microtomes for TEM specimen preparation. Indeed, resort to a cryo-ultramicrotome for a 1 μm section of a difficult polymer may be the only possible solution to a problem. However, such equipment is expensive and restrictive on specimen size so is not a particularly satisfactory solution for light microscopy preparation.

Electrical cooling devices based on the Peltier effect deserve mention. Although the lowest temperatures reached by commercially available cooling stages of this design are not usually low enough to effect substantial improvement to cutting characteristics of polymers, they do have the advantage of offering high freezing rates, high stability, controllability, and the ability to hold difficult specimens. If the intention is to use a microtome freezing stage just to embed a specimen in ice as mentioned earlier, then these are useful devices and with luck some improvement with difficult materials might be experienced.

Some of the polymers, which are difficult to thin-section because they are too hard or tough and reside to the right in Fig. 2.1, can sometimes be sectioned if their temperature is *raised*. Obviously care must be taken not to modify the morphological features under investigation.

A rather primitive but often effective method of providing the necessary heat is to blow air from a domestic hair dryer over the specimen. This has been found particularly effective when sectioning high-molecular-weight poly(methylmethacrylate). Better control and a wider range of temperatures can be achieved by the use of a specially constructed enclosure for the specimen and microtome knife, and an industrial air heater. To the best of the author's knowledge, specimen and knife heaters capable of reaching the temperatures required are not supplied as standard equipment by microtome or accessory manufacturers.

Mounting sections

Sections may be mounted between microscope slides and coverslips using the methods established for biological work. The choice of mounting medium is partly dictated by the polymer, since great care must be taken to avoid solvent attack which may take the form of complete solvation or swelling of the microstructure.

Table 2.1. *The chemical resistance of some common polymers (1 = resistant; 5 = not resistant)*

Polymer	Dilute acids	Concentrated acids	Dilute alkalis	Concentrated alkalis	Aliphatic hydrocarbons	Alcohols	Ethers	Esters	Ketones	Amines	Aromatics	Oxidizing agents
Low-density polyethylene	1	1	1	1	4	3	5	3	3	–	5	5
High-density polyethylene	1	1	1	1	3	1	3	1	1	–	2	5
Polystryrene	1	2	2	1	4	1	4	5	5	1	4	3
Acrylonitrile butadiene styrene copolymer	1	2	1	1	3	2	5	5	5	1	5	4
Poly(vinyl chloride) (unplasticized)	1	1	2	1	2	4	1	5	5	2	4	2
Poly(tetrafluoroethylene)	1	1	1	1	1	1	1	1	1	1	1	1
Poly(methyl methacrylate)	2	4	1	1	2	3	2	5	4	1	4	3
Nylon 6	5	5	2	1	2	2	1	1	1	1	2	5
Nylon 6:6	5	5	2	1	2	2	1	1	1	1	2	5
Polysulphone	1	1	1	1	2	1	–	–	5	–	5	1
Phenol formaldehyde resin	3	5	3	5	–	2	2	2	2	–	2	–
Urea formaldehyde resin	4	5	3	5	–	2	1	2	2	–	2	–
Melamine formaldehyde resin	3	5	2	5	–	1	2	2	2	–	2	–
Epoxy resin	2	4	2	2	–	1	1	4	2	2	4	5
Polyurethane rubber	1	4	1	1	2	4	3	5	5	–	4	1
Cellulose acetate	1	5	3	5	1	3	1	5	5	4	2	5
Polypropylene	1	2	1	1	2	1	3	2	2	1	4	5
Poly(4-methyl pentene-1)	1	1	1	1	5	2	5	4	3	–	4	4
Ethylene vinyl acetate copolymer	1	3	1	1	4	2	4	4	5	2	–	4
Polyformaldehyde	2	5	1	1	1	1	1	2	1	–	2	4

Table 2.1 shows the chemical resistance of a selection of common plastics and rubbers. It will be seen that certain materials (e.g. polystyrene) are particularly sensitive to solvent attack. Even where moderate resistance in indicated, traces of solvents in mountants can produce long-term modifications of the polymer especially if it is of low molecular mass. The problem of preparing mounts for long-term use is therefore quite severe, especially when it is also recognized that for most polarized-light and interference microscopy work a reasonably close refractive index match between specimen and mountant is required. In practice long-term mounts are usually prepared using xylene-free Canada Balsam or Euparal and any refractive-index mismatch is tolerated.

Shorter-term mounts (for use only over several days or weeks) are prepared using a variety of organic liquids, the choice of which depends on the refractive index (mean index if anisotropic) of the specimen and its chemical resistance. Table 2.2 shows the refractive indices of a number of common polymers and Table 2.3 shows typical liquids used for mounting some of them.

Table 2.2. *The refractive index of selected polymers*

Material	Refractive index (average value n_D^{20})	
Poly(tetrafluoroethylene)	1.35–1.38*	
Tetrafluoroethylene hexafluoropropylene copolymer	1.34	
Poly(vinylidene fluoride)	1.42	
Poly(vinyl fluoride)	1.46	
Poly(butyl acetate)	1.463	
Poly(ethyl acrylate)	1.468	
Poly(vinyl acetate)	1.47–1.55	(according to acetate content)
Poly(methyl acrylate)	1.48	
Poly(4-methyl pentene-1)	1.47*	
Polyformaldehyde	1.47–1.48*	
Poly(methyl methacrylate)	1.49 (5)	(cast sheet and moulding compound)
Polypropylene	1.49–1.50	
Polyisobutylene	1.505–1.510	
Polyethylene (LD)	1.51*	
Polyethylene (HD)	1.53*	
Polyisoprene (Nat. rubber)	1.52	
Polyacrylonitrile	1.52	
Polybutadiene	1.52	
Styrene acrylonitrile copolymer	1.57	
Nylon 6	1.53*	
Nylon 6:6	1.53*	
Nylon 11	1.52*	
Cellulose acetate	1.50	
Poly(vinyl chloride)	1.54	
Neoprene rubber	1.55	
Epoxy resins	1.57–1.61	(depends on composition)
Polystyrene	1.59	
Poly(ethylene terephthalate) (film)	1.60–1.63	
Poly(ethylene terephthalate) (amorphous)	1.58	
Poly(ether sulphone)	1.65	
Poly(aryl sulphone)	1.67	
Polyimide	1.70	
Polycarbonate	1.58	
Ionomers	1.51	

*Average values. Actual value will depend on degree of crystallinity.

Mixtures of glycerol and water or glycerol, potassium mercuric iodide and water are suitable for use with polymers if a continuous range of refractive indices is required. To a satisfactorily close approximation, the refractive index n of a mixture of two miscible liquids of indices n_1 and n_2 is given by

$$(v_1 + v_2)n = v_1 n_1 + v_2 n_2$$

where v_1 and v_2 are the volumes of the two liquids used.

Commercially available sets of liquids with incremental steps of index used for specimen refractive-index and optical-dispersion measurements should be used with some caution. These liquids are mainly mixtures of aliphatic and aromatic hydrocarbons and can attack the specimen. Such interaction may not be immediately

Table 2.3. *Some mounting media for common polymers*

Polymer	Liquid	Refractive index (n_D^{20}) of liquid
Polythene (HD, LLD and LD))		
Polypropylene)	Dimethyl phthalate	1.51
Polybutene)		
Poly(vinyl chloride)	50/50 v/v Clove oil/ Aniseed oil	1.55*
Nylon (6, 6.6, 6:10)	Clove oil	1.54
Poly(tetrafluoroethylene)	(Acetone	1.36
	(Perfluorokerosene	1.30
Poly(ethylene terephthalate) film	Cassia oil	1.61*
Poly(methyl methacrylate)	Dibutyl phthalate	1.49*
Poly(4-methyl pentene-1)	Olive oil or glycerol	1.47
Polystyrene	Cassia oil	1.61*
Poly(tetramethylene terephthalate)	Quinoline	1.62
Polycarbonate	Aniline	1.58
Natural rubber (Polyisoprene)	Cedarwood oil	1.52
Neoprene rubber	Aniseed oil	1.55
Poly(ether sulphone)	Monobromonaphthalene	1.66
Poly(aryl sulphone)		
Polyformaldehyde	Olive oil	1.47

*Some signs of attack after a short period of immersion, particularly of low-molecular-mass materials.

obvious in the short term, but tell-tale signs are a loss of birefringence of the section, swelling, buckling, or a loss of relief as the section and liquid become closer in refractive index.

Staining

At the very outset of specimen preparation the source of the necessary contrast in the final microscope image should be under consideration. In view of the obvious parallel between the preparation of synthetic and natural polymers, the selective staining of sections might be considered as a route to contrast development or enhancement. Unfortunately, the chemical structure of synthetics does not generally lend itself to this approach; in practice it is difficult to achieve sufficient coloration for the technique to be useful. As a consequence, polymer microscopy depends more heavily on optical methods for obtaining contrast and identifying phases.

2.4. Melt pressing and solvent casting

As an alternative to thin sectioning with a microtome, thin films of thermoplastic may be produced by pressing small pieces of material between a slide and coverslip on a hotplate at a temperature at which the polymer flows easily, then allowing the preparation to cool to room temperature. Such preparations are useful for observing modifications to crystalline texture produced by different thermal treatments and for determining the nature and particle size of certain additives (e.g. pigments). The method has some serious limitations however.

Table 2.4. *Recommended melt pressing temperatures*

Polymer	Temperature (°C)
HD and LLD polyethylene	170
LD polyethylene	150
Polypropylene and propylene/ethylene copolymers	200
Nylon 6	250
Nylon 6:6	280
Nylon 11	220
Poly(ethylene terephthalate)	290
Polyformaldehyde	190
Polystyrene	130
Poly(methyl methacrylate) (moulding grades)	260
Acrylonitrile/butadiene/styrene terpolymer	160
Poly(vinyl chloride)	190
Styrene/butadiene/styrene terpolymer	160
Polycarbonate	260
Poly(4-methyl pentene-1)	275
Poly(tetrafluoroethylene)	360
Poly(ether sulphone)	350
Poly(butylene terephthalate)	250

1. The distribution of the phases in a composite material will probably be changed.

2. Degradation may occur giving rise to non-representative crystalline texture, particularly round the edges of the preparation.

3. The crystalline texture developed will not necessarily be that which would be developed in bulk under equivalent thermal conditions due, mainly, to surface nucleation and melt orientation.

Despite these disadvantages, melt pressing is a quick, cheap, and therefore popular technique. Pressings should never be thicker than about 30 μm, which may mean initially using a piece of the material no larger than a pin head. A useful practical tip is to invert the preparation on the hotplate once the polymer has melted, so that pressure is subsequently applied to the slide rather than the coverslip. A simple tool for pressing on the slide or coverslip can be made from a glass rob by melting its end in a flame and pressing it against a suitable flat surface of low thermal conductivity until cool.

As mentioned above, the effect of different thermal treatments can readily be investigated by this preparative method. In such an investigation it will probably be necessary to rapidly cool some specimens by transferring them directly from the hotplate into iced water or an alternative coolant such as an ethanol/solid CO_2 mixture. The thermal shock on a standard slide and coverslip usually leaves the would-be microscopist with little left to look at except broken glass! Silica slides and coverslips overcome this problem, but at a price.

The recommended pressing temperatures for a number of polymers are shown in Table 2.4. They may need to be adjusted upwards or downwards to cope with materials of exceptionally high or low molecular masses. Materials now shown in the table should be pressed about 30°C above their crystalline melting points or softening temperatures as appropriate.

2.5. Preparation of hard or brittle materials

As outlined earlier, some polymers (e.g. most thermosetting resins) or highly filled materials, cannot be successfully sectioned by microtomy and other methods of displaying their internal texture must be employed.

Standard methods of metallographic specimen preparation may be used. The polishing rates for polymers are, however, much faster and unless large numbers of specimens are to be examined, manual as distinct from automatic methods are acceptable. Certain precautions are necessary.

1. Polishing rates should be controlled to avoid excessive heating, especially when preparing thermoplastics.
2. Guard against chemical attack of the polymer by lubricants. Water containing a little detergent is sometimes the only safe liquid.
3. The tendency of small specimens to rapidly produce a rounded profile due to higher polishing rates at the edge can be countered by encapsulation in an epoxy or acrylic resin or simply by mounting the specimen in a suitable metal jig.

Diamond compounds are excellent abrasives for the harder polymers, particularly the thermosets. For softer materials alumina and iron oxide are used. These are dispersed on a suitable cloth (a Selvyt cloth is usually very satisfactory) using a generous application of lubricant. Typically a polymer will be successfully polished with 15 μm, 3 μm, then 1 or 0.25 μm abrasive.

Fig. 2.4 shows the surface of glass-fibre-reinforced polypropylene at different stages of polishing. The visibility of the fibre cross-sections is acceptable only after the final stage of polishing. Contrast is developed from the difference in reflectivity between the glass and the polypropylene. A common problem arises if there are great local variations in hardness of the specimen. For example, phenol formaldehyde resin frequently contains a high loading of cellulosic material and this traps the coarser grades of abrasive which are then inconveniently released in the later stages of polishing. This is difficult to overcome completely, but thorough washing of the surface, preferably using an ultrasonic bath, minimizes the problem. In cases where the difficulty is caused by voids or holes in the specimen, these can usefully be filled with epoxy resin, or similar, prior to the start of polishing.

The trapping of abrasive by softer polymer phases in a hard matrix composite may be overcome by low-temperature polishing (below the T_g of the soft phase). A dish of liquid nitrogen below the plate supporting the polishing cloth may be all that is required.

Micromilling techniques have been shown to be applicable to the harder polymers and can yield excellent results.

Etchants for prepared surfaces may be selected on the basis of the information in Table 2.1. As well as selective removal of material from a prepared surface by solvation it is also possible to use selective solvent swelling to generate surface relief. Often an exposure to solvent vapour for a few minutes is sufficient.

Specimen preparation 23

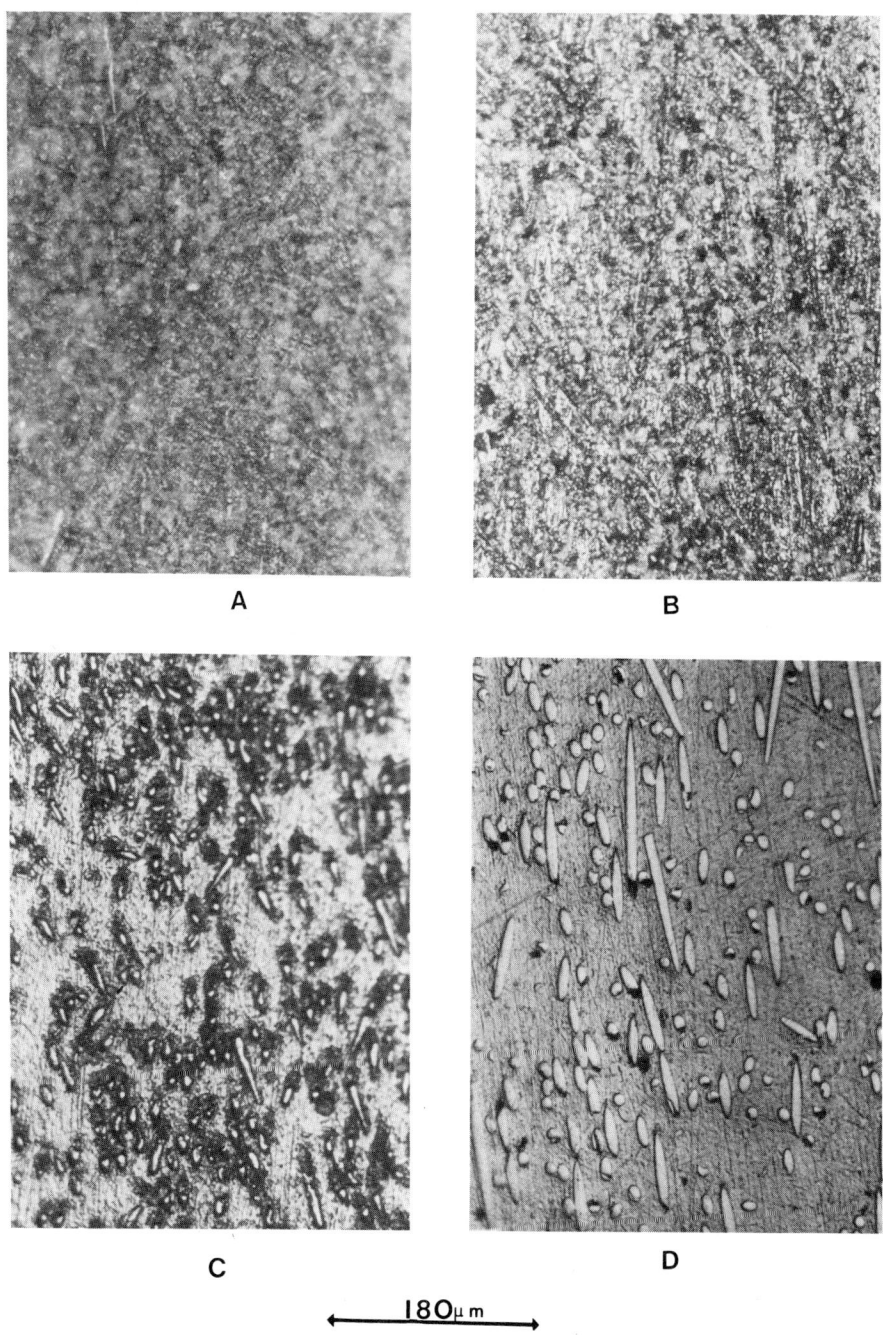

Fig. 2.4. Stages (A–D) in the polishing of polypropylene containing glass fibres (reflected light).

Obtaining thin sections from hard materials can be attempted using standard petrographic methods and the abrasives and techniques described above probably preceded by preliminary thinning using a series of wet carborundum papers. In this technique a polished surface is first produced on one side of a thin sawn slab of the material. This surface is then struck to the slide with a suitable adhesive before grinding and polishing away the polymer from the exposed upper surface. The main problem with the method applied to polymers is the adhesive. Again care is taken not to modify the material by heat or solvent attack. The cyanoacrylate *superglues* have been found valuable in some cases.

2.6. Preparation for surface studies

It will be seen in Chapter 3 that considerable information is available from surface studies on polymers, particularly on films. Surfaces may be examined directly using reflected light or via replicas. The advantages of replication are:

1. The original surface is left intact and may, if necessary, be reexamined subsequently for changes.

2. The replicating material may pose fewer problems of specimen preparation than the specimen itself.

3. Replicas are generally thin and transparent enough to allow transmitted light observation of surface roughness.

Replicating materials include methyl cellulose or gelatin dissolved in water, or polystyrene beads dissolved in benzene. The concentration depends on the viscosity required but is generally in the range 1–5 per cent w/v of the solid. A small amount of detergent may be added to the water-based solutions to aid wetting of the polymer surface. After lightly coating an area of the surface to be examined with the solution, the water or solvent is allowed to evaporate off (or gently encouraged to do so with an i.r. lamp) and the remaining thin film is eased off the polymer surface with forceps.

For transmitted-light work the replica needs to be shadowed using techniques familiar to electron microscopists. A metal, conveniently aluminium, is evaporated from a point source on to the surface at an angle to the vertical. The angle, and the amount of aluminium deposited, may be varied according to the height of the surface structures but 14°, giving a shadow-to-height ratio of four, is typical. The prepared replica is then mounted between slide and coverslip using a mounting fluid of precisely the same refractive index as the replicating material. The image formed in the transmitted-light microscope then reveals only the distribution of aluminium which in turn is related directly to surface roughness. Measurements of roughness are possible by shadow-length measurement.

The surfaces of thin films of polymer, such as are used for packaging purposes, may be examined directly in transmitted light using the same shadowing and immersion technique.

The examination of polymer surfaces by reflected light avoids the use of replicas but poses its own problems. The percentage reflectivity R of polymers at normal incidence may (since they are dielectrics) be calculated from the expression

$$R = 100 \left(\frac{n-1}{n+1}\right)^2.$$

Inserting figures for polytetrafluoroethylene and polyether sulphone from Table 2.2 shows that the extremes of reflectivity for polymers are 2.2 and 6.7 per cent, respectively. Such low reflectivities give poor image brightness and light scattering from within the specimen (which is usually transparent or translucent), which is a problem since its intensity may be comparable with that coming from the surface. For this reason most specimens are metal-coated to increase the reflectivity. Traditionally this involves sputter coating with gold, although the choice of metal is not critical. As with the evaporation technique mentioned above, overheating of the specimen during coating must be avoided. The thickness of the metal layer is also not critical, a few tens of nanometres being sufficient to substantially increase reflectivity.

3

Surface studies

3.1. Introduction

The surfaces of plastics products are important from both aesthetic and technical viewpoints. Surface-roughness differences may produce apparent differences in the colour of pigmented products, changes in mechanical strength, and differences in frictional characteristics. An examination of the surface of such products is therefore of clear practical importance. It is convenient to look at some typical surface structures and their interpretation under three headings: first, the surfaces of mouldings and extrudates; second, fracture surfaces an examination of which may help to isolate the cause of mechanical failure; third, the more specialized area of film surfaces. A close look at the latter is justified on the grounds that the surface properties of films are of special importance.

3.2. The surface of mouldings and extrudates

In injection moulding, molten polymer is injected into the mould under a high pressure which is held until solidification takes place. The surface of most commercial mouldings will therefore show replication of the mould surface. Common incident-light examination usually shows little detail but incident-light *differential interference contrast* (DIC) microscopy carried out on metal-coated samples is more revealing. Fig. 3.1 shows a typical moulding surface looked at in this way. The replication polishing marks on the mould are clearly visible. The rounded 'volcano-like' structures are due to additive particles in the polymer. Note that the DIC technique greatly enhances the apparent surface roughness and may present a pessimistic picture of a mould surface.

Local absence of mould replication may be due to

1. The contraction of the material away from the mould before solidification;
2. Insufficient moulding pressure;
3. The build-up of additives or degraded polymer on the mould surface.

The loss of mould-surface detail in case 2 may not be complete since contraction will normally take place late in the mould cooling cycle, after replication has taken place but before relaxation of the surface becomes impossible. Fig. 3.2 shows the effect of low mould pressure (in this case a blow moulding). There are local areas where spherulic structures (see Chapter 4) are visible at the surface and mould

Surface studies 27

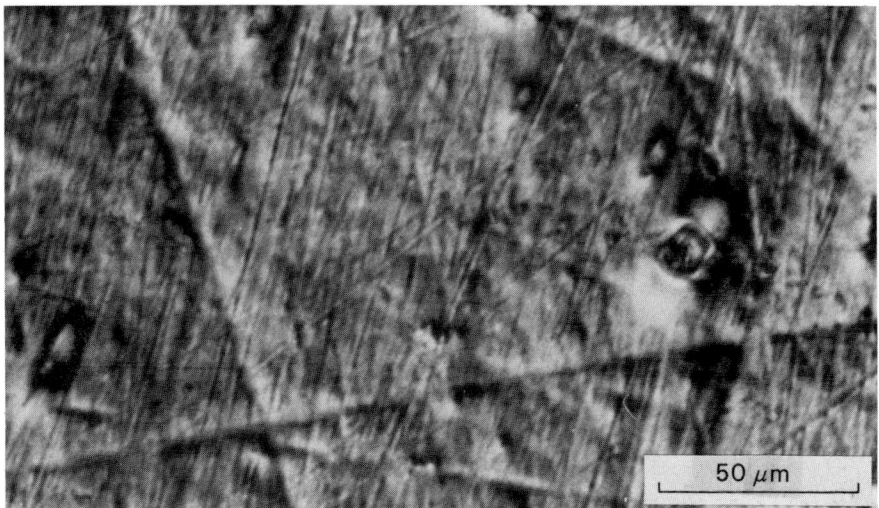

Fig. 3.1. Surface of poly acetal moulding (reflected light DIC).

replication is absent. Such a surface is described as a *free surface* and indicates poor mould—polymer contact.

The surfaces of blow mouldings are usually rather different from those of injection mouldings in that the surface will contain a memory of the surface of the extrudate which is subsequently inflated in the mould. This is clearly evident as the bands of rougher texture shown in Fig. 3.3.

The surface roughness of injection mouldings in the vicinity of the injection point (the *gate*) is of particular relevance since this area is often intrinsically weak, and surface defects will be particularly disadvantageous.

Dark-field microscopy is of considerable use on mouldings if subsurface information is required. Thus the technique may be used on uncoated transparent mouldings which contain pigments or other light scattering entities to establish their character and distribution. The advantage lies in the fact that no specularly reflected light from the comparatively smooth moulding surface takes part in image formation. The information contained in the image then refers only to light scattered on and just under, the surface. Obviously if the surface contains scratches or other surface damage, these will also be well imaged and may cuase confusion in image interpretation. Nevertheless, on account of the extreme sensitivity of the technique to small inhomogeneities, it is of considerable value to the polymer microscopist.

Reflected polarized light may also be used profitably, but again for examining additives rather than the polymer itself. Suitable adjustment of the polarizer and analyser of the microscope can achieve a situation when almost no light returns from the polymer surface itself, but only from entities changing the state of polarization by double refraction or scattering. The image often bears a similarity to the dark-field image, and is again achieved using uncoated specimens.

Textured surfaces on mouldings, such as might be used to produce a simulated

28 Light microscopy of synthetic polymers

Fig. 3.2. Surface of polypropylene moulded at low pressure (reflected light). (A) Mould-replication area; (B) free surface area.

leather look, are almost too rough to be examined with a light microscope. A low-power stereomicroscope is useful for texture of lateral extent above a few tens of micrometres but finer scale structure is better examined in the SEM.

Extrudate surfaces are 'free' surfaces since solidification will take place not against a solid but a fluid (usually water or air). In this case the surface more closely reflects the internal texture of the material. However, the process is normally designed to cool the surface very rapidly so what is seen at the surface in terms of crystallization texture may not reflect the texture of the bulk of the product. However, as will be illustrated when discussing polymer films below, the surface of an extrudate may also be modified by

Fig. 3.3. Surface of blow-moulded polyethylene (reflected light).

1. Imperfections in the extruder die. These will usually give rise to linear surface defects varying in width from a few micrometres to millimetres;

2. Rheological effects, which in their least evident form may be detectable only with the DIC method.

At the other extreme, severe distortion of the extrudate may take place making anything other than simple macrography of the specimen impossible.

3.3. Fracture surfaces

There are two aspects to fracture-surface examination. On one hand there are *in-service* fractures over which the microscopist has no control and which generally are inconveniently irregular in profile! The other type of fracture is that produced as part of a controlled specimen preparation routine. This may have been undertaken because the specimen cannot be sectioned and does not lend itself to polishing. The location of the fracture can be predetermined, at least approximately, by notching the specimen and an appropriate temperature chosen to give a fairly smooth fracture surface. Materials normally ductile at room temperature, such as polyethylene, may need immersion in liquid nitrogen prior to being snapped.

For the first type of surface the problem is usually to characterize the type of fracture and to locate its initiation point. In the second type, the objective is

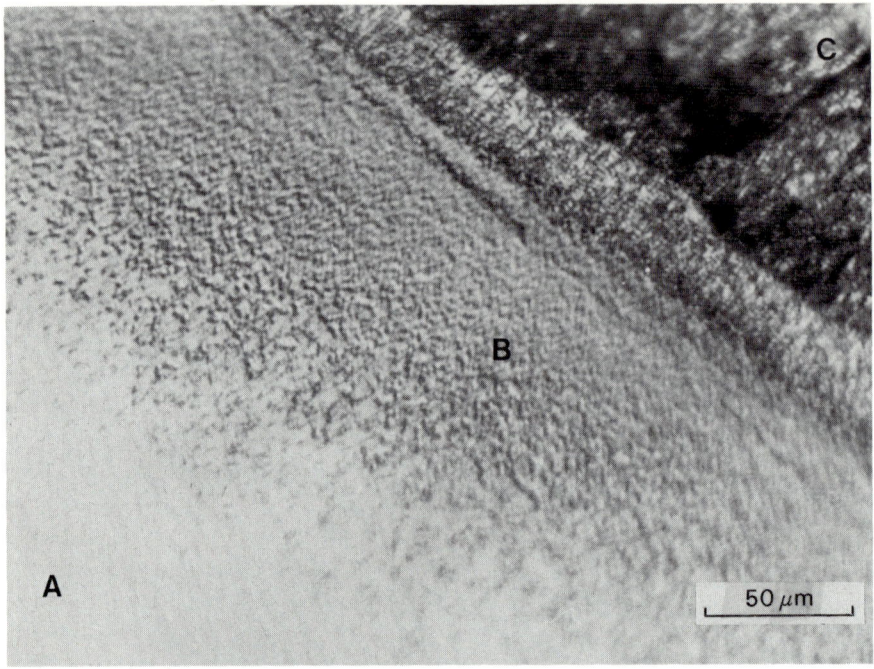

Fig. 3.4. Typical brittle failure of polystyrene (reflected light DIC). (A) *Mirror* region; (B) *mist* region; (C) *hackly* region.

normally to examine the revealed internal structure of the specimen. In either case it is good policy to examine both halves of the fracture.

Fracture surfaces display structures of such variety that almost all incident-light techniques (and some transmission techniques, if replicas are employed) have a part to play. However, two particularly difficult types of specimen are

1. High-speed, forking fractures which give a very rough irregular surface;
2. Highly ductile fractures in which considerable deformation has occurred, accompanied by 'stress whitening' (due to the development of microscopic cracks and voids) and high molecular orientation.

In both cases the use of an SEM rather than a light microscope is recommended. Fortunately, in practice, many 'accidental' fractures and almost all 'preparation' fractures are not in this category and the light microscope is applicable.

Slow crack-growth regions of polymers tend to have mirror-like surfaces. DIC incident-light methods are both highly effective and easy to apply to such specimens. Common light methods on the other hand, yield little information.

A full description of polymer-fracture morphology is beyond the scope of this text and is adequately covered elsewhere. However two illustrations follow which demonstrate the effective use of the light microscope.

Figure 3.4. shows a typical brittle fracture surface. Three zones, often described

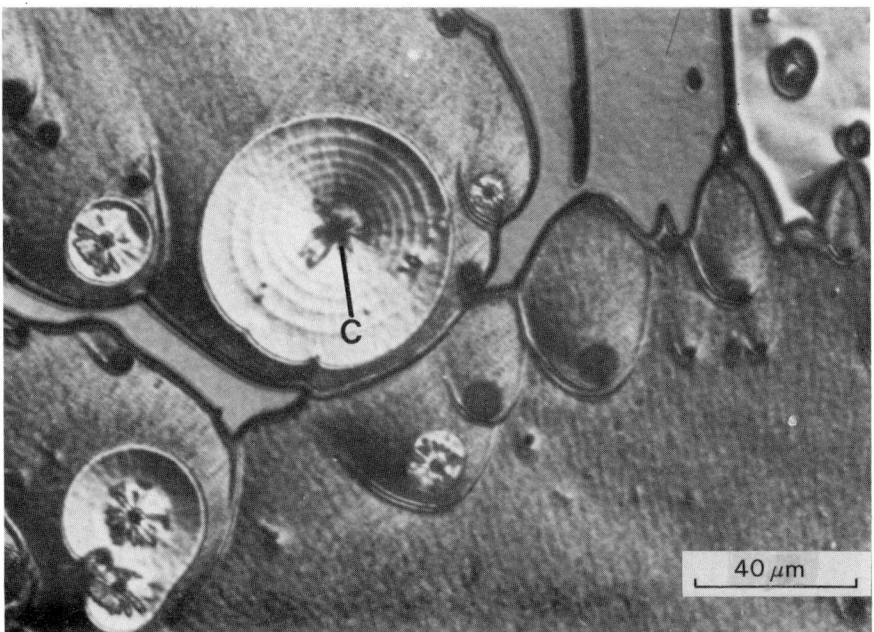

Fig. 3.5. Fracture surface of contaminated polystyrene (Reflected Light DIC). (C) Contamination.

as the *mirror, mist,* and *hackly* regions, are shown. DIC reveals considerbly more information about such surfaces than any other technique, although dark-field microscopy will detect some of the features present. The presence of these zones, their shape and size, allows the type of fracture to be characterized in terms of the type of loading and the speed of fracture propagation. They also assist with locating the origin of the fracture which is usually at the centre of concentric markings.

Fig. 3.5 shows the fracture surface of a broken polystyrene bar. A number of concentric structures are visible which indicate fractures starting from a number of centres. A close examination of these structures reveals a particle of contamination at the centre of each. Such a particle acts as a stress concentrator and leads to premature failure of the moulded component.

3.4. The surfaces of films

The surface microstructure of a plastics film arises from the method of manufacture (e.g. melt extrusion, solvent casting), the rheological properties of the melt or polymer solution, and the presence of additives. Film may be manufactured by a number of different production techniques but the majority involve extrusion of polymer melt either into the form of a tube or on to a casting drum. Drawing of the film to induce the desired molecular orientation may take place either during extrusion or subsequently, often after re-heating the extruded material. Drawing is

32 *Light microscopy of synthetic polymers*

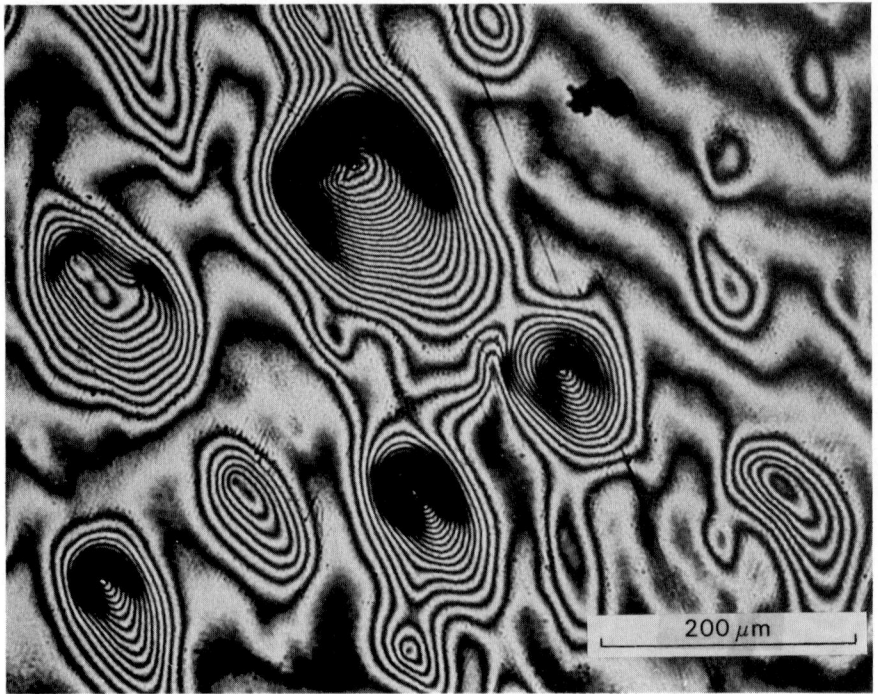

Fig. 3.6. Surface of polypropylene film (reflected-light two-beam interferometry).

usually done in two directions; sometimes simultaneously, sometimes sequentially. Films produced from solutions of polymers in solvent are usually made by a casting process, again with subsequent drawing.

Some of the microsctructural features exhibited by the melt extrustion of tube route are described below. All may be examined by the direct-shadowing and transmitted-light method described in Chapter 2, although the shadowing angle will need to be adjusted to give an appropriate shadow structure height ratio for the different scales of structure. The incident-light DIC method may be used to advantage on the crystallization structures, and the more shallow of the extrusion defects. The remainder are often too steep-sided for interpretable imageing by DIC and are best best examined by a technique having less vertical sensitivity.

Incident-light interference methods, either two-beam (e.g. the Mirau System) or multiple-beam, can be used to quantify surface roughness. Fig. 3.6 shows the surface of polypropylene film imaged with a two-beam system and monochromatic light. The multiple-beam method is generally excessively sensitive for all but the smoothest of surfaces and when precise surface contouring is required. For these techniques it is essential that the film is metallized to maximize reflectivity.

 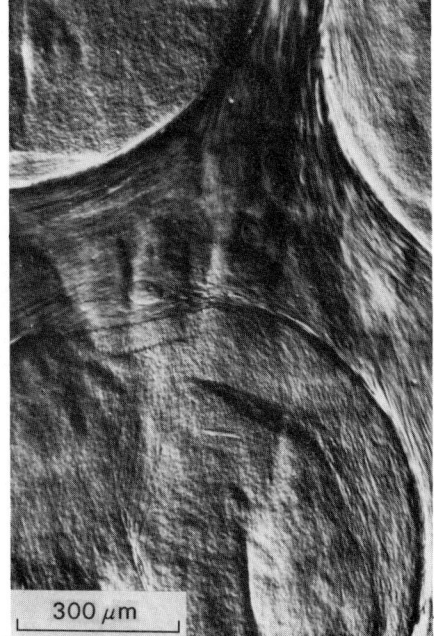

Fig. 3.7. Crystallization structures on polyethylene film (transmitted-light shadowed specimen).

Fig. 3.8. Rings on polypropylene film (reflected-light DIC).

Typical extruded tubular film structures

Crystallization structures

Surface crystallization is not influenced by an adjacent solid as in moulding. On the other hand the nucleation and growth of crystalline texture can be strongly influenced by the extrusion and drawing conditions, by the rapid rate of cooling, and by the shear applied to the polymer during extrusion and at the die lips. Generally very fine crystalline texture results. On the polyethylene film shown in Fig. 3.7 for example, crystallization nuclei only $\sim 0.5\,\mu m$ across can be seen on the surface. Coarser structures can occur, but rarely are identifiable spherulites visible. In rare instances where large surface spherulites do occur then pits can develop at the points where three or more spherulites meet. The origins of these pits appear to be similar to those of interspherulitic voids in the bulk (see Chapter 4). If a material exhibiting such pits is reheated and biaxially drawn (as is the case for some polypropylene films), large shallow surface rings may result. Some rings of this type are shown in Fig. 3.8.

Extrusion defects

Apart from the obvious continuous longitudinal structures which can be attributed to disruption of the surface by imperfections or deposits on the die lips ('die lines')

Fig. 3.9. Extrusion structures on polyethylene film (transmitted light-shadowed specimen).

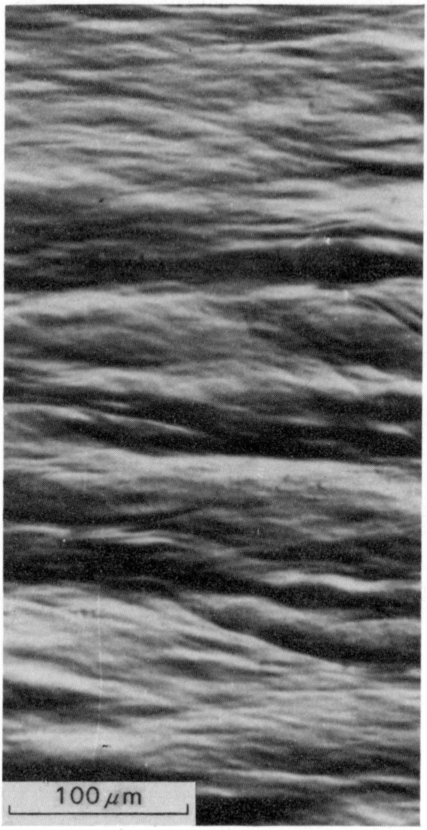

Fig. 3.10. *Sharkskin* structures on polyethylene film (transmitted light).

surface structures can arise from the strains in the extrudate or film, produced during extrusion. Extrusion through a die and the simultaneous application of haul-off to the melt can give rise to surface structures up to 5 μm high and up to 100 μm across such as those in Fig. 3.9. The formation of these structures is often associated with the presence of a fraction of high-molecular-mass material and is thought to be due to a buckling of the surface by differential strains across the extruded profile.

Note that the azimuth effect characteristic of the DIC system may be used to reduce the visibility of the lines on the surface of the film by rotating it until they are in the shear direction of the optical system. More equidimensional features remain visible. Gross viscoelastic defects can also occur due to melt fracture. This is often referred to as *sharkskin* and is usually visible to the unaided eye. Typically the structure consists of sharp ridges roughly at right angles to the extrusion or *machine* direction (Fig. 3.10).

Effects produced by additives

Particulate additives, contamination, or particles of cross-linked polymer can give rise to the disruption of film or extrudate surfaces. Apart from surface bumps of obvious origin, craters or shallow depressions can be produced if the particles, or clusters of particles, pull apart as a result of uniaxial or biaxial drawing.

Non-particulate additives can *bleed* or *plate out* on to a surface during or after film manufacture and submerge structures which would otherwise be easily visible. Incident-light optical systems with a high vertical resolution such as DIC are necessary to detect thin *plate out* layers directly but additives which crystallize on reaching the surface are best seen using reflected plane-polarized light.

Surface damage

The comparative softness of film polymers means that their surfaces are readily scuffed and scratched. Any film that has been handled or subjected to mechanical abrasion will show an abundance of surface scratches.

Typical cast film structures

The microstructural features visible on extruded tubular films may also be visible but to a lesser degree, on cast film. Thus the presence of particulate additives will still be revealed by surface excrescences. However, the casting operation complicates interpretation because

1. One side of the film will normally be cooled faster than the other. Surface textural differences will result, particularly in terms of crystallization structures, and these may still be evident in the finished film.

2. There will be replication of the metal casting drum surface. To prove whether or not a given film surface feature is the result of replication during film manufacture, it is sometimes useful to replicate an area of the drum as described in Chapter 2. Note, however, that the finished film will have been drawn, and drum replication features will have been stretched in the plane of the film in either one or two directions.

3. Extrusion structures are usually much less in evidence in the case of melt cast films and absent for solvent cast.

3.5. Pitfalls and precautions

Although in practical terms films represent almost ideal specimens for microscopical examination, there are a number of pitfalls to be avoided and precautions to take. Failure to heed these, as the author knows well, can lead to incorrect observations and considerable embarrassment!

1. When sampling film, mark the two surfaces for easy identification. One surface can look very like another, yet knowing which is the 'inside' or 'outside' of

the film may be vital. Also mark the machine (MD) direction for referencing any elongated structures.

2. A film may contain a mobile fluid or semifluid constituent such as an antistatic agent. This can prevent proper deposition of metal on the surface. The imperfect deposited layer normally exhibits a network of cracks which it is possible to interpret as genuine film microstructure.

3. Overheating of film during metallizing is remarkably easy. Because the film may contain highly oriented molecules, it can shrink and modify its surface microstructure with only modest heating. It is not necessary to actually melt the polymer for 'relaxation' to take place. If in doubt, carry out a preliminary observation on unmetallized film. The contrast will be poor but some idea of what to expect after metallizing will be gained. Ink from a black felt-tipped pen put on the reverse side or underside of the film will help minimize reflections from this surface and increase image contrast. However, make sure the microscope system contains a heat filter otherwise 'beam damage' can occur as the black layer absorbs infrared radiation very effectively!

4. When using DIC to examine a film surface be sure to examine it from the metallized side. Attempting to see it through the polymer layer can give puzzling results since the film will almost certainly be optically anisotropic and this modifies the image produced by the DIC system which uses polarized light.

5. Be sure to rotate the specimen when using DIC. Many surface structures will be elongated as the result of the drawing operation. Look for the position of best visibility and bear in mind that features in a different orientation will be supressed.

4

Observation of crystalline texture

4.1. Introduction

Despite the long-chain molecular structure of synthetic polymers, many of these materials are capable of crystallization. Unlike many other materials however, the amount, or degree of crystallinity achievable may be substantially less than 100 per cent, even under favourable crystal-growth conditions. Often in the production of manufactured articles, the growth conditions are far from favourable and a crystallinity below 20 per cent can result.

The factors influencing the intrinsic capability of a material to crystallize and the typical morphological features produced have been outlined in Chapter 1. The main and most informative feature visible in the light microscope is the spherulite. The actual appearance of spherulites in a manufactured product will depend on a number of variables. Clearly one of these will be the choice of observation or contrast mode. Although observation between crossed polars in a polarizing microscope might be considered the normal method of viewing crystalline texture, other modes are sometimes more useful. These are discussed later, but for the moment we will consider the factors influencing the appearance of spherulitic material between crossed polars. Some typical polyethylene spherulites are shown in Fig. 4.1. To describe why they have the appearance shown some further comments on the structure of a typical polymer spherulite are necessary.

4.2. Structure and optical nature of a spherulite

Consider first the observation of a single lamella (see Chapter 1) between crossed polars. The geometry will normally dictate how the lamella rests on the microscope slide and thus the direction in which we look at it. Between crossed polars the lamella is invisible, i.e. the field of view remains dark. Does this mean that even crystalline polymers are optically isotropic and not doubly refracting or *birefringent*? Fortunately for the polymer microscopist this is not the case. In the experiment described we will be looking at the lamella, or single crystal, down, or nearly down, its optic axis. Polarized-light theory indicates that in such a situation double refraction will not be seen. An extension to our imaginary experiment in which the lamella is tilted to be examined 'side on' would reveal double

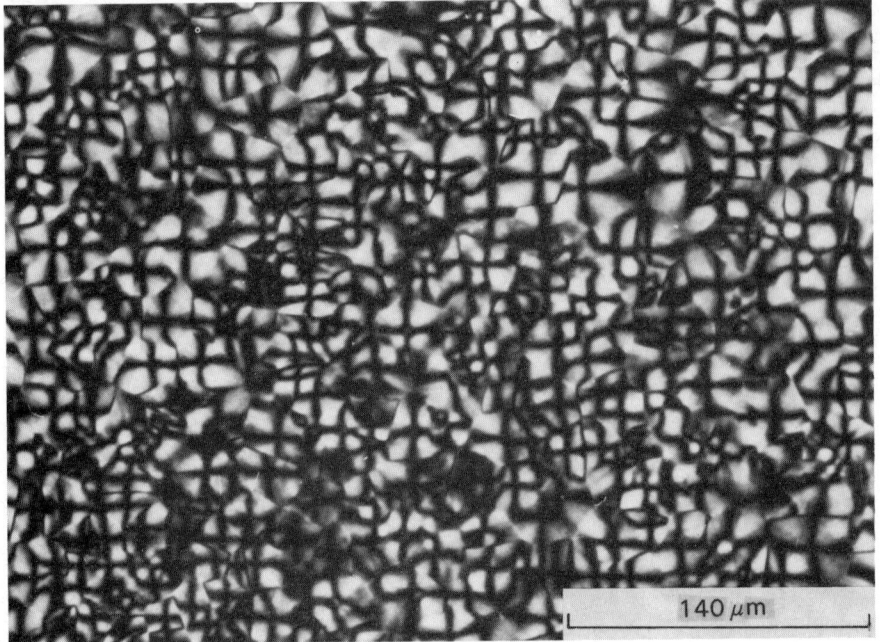

Fig. 4.1. Well-developed polyethylene spherulites (crossed polars).

refraction effects, although in practice such an experiment is made difficult by the extreme thinness of the lamella.

The optical character of a single lamella is determined both by the arrangement of the molecules and by the optical characteristics of the molecules themselves. As outlined earlier there is a regular packing of molecules organized perpendicular to the plane of the lamella. Indeed, some regularity of arrangement is implied by describing the material as being crystalline. There is, however, considerable evidence that many imperfections of arrangement occur (many more than resulting from dislocations or other crystal defects in more traditional crystalline materials). The other factor determining the optical character of a single lamella is the manner of response of the actual molecules to the light wave passing over it. With just a few rare exceptions, polymer molecules are optically anisotropic in that they respond differently for light waves passing over them vibrating in different directions. This means that when organized into regular arrangements the polymer will then exhibit different refractive indices in different directions; that is, single crystals or lamellae will themselves be optically anisotropic, exhibiting double refraction effects. The exceptions mentioned above (poly(4-methyl pentene-1) at room temperature is a good example) are interesting in that they demonstrate that *both* organization and molecular anisotropy are necessary for the bulk material to be optically doubly refracting. In these unusual cases the required organization exists within the lamella of these semicrystalline materials but the molecules happen to be, on account of their atomic structure, optically isotropic. The lamellae are therefore not doubly refracting or birefringent. In all cases the optical character of the lamellae is carried over to the larger-scale morphological features of the material as described below and so dictates how it will appear in the polarizing microscope.

Electron microscopy studies on the crystallization behaviour of polymers from the melt have shown that they crystallize in a spherulitic form with spherulite growth taking place outwards from a central nucleus. Often a radiating and branching fibrilar structure is shown and there is abundant evidence that these are composed of the lamella-type structures described above. This implies that the fibrilar structures, and thus the spherulites themselves, will be optically anisotropic and as such, visible in the polarizing microscope. This proves to be the case and often both the individual spherulite boundaries and the contained fibrils are clearly visible with the resolution available.

Optical and X-ray diffraction data agree that, perhaps surprisingly, individual molecules are arranged perpendicular to the radius of a spherulite. This does not, however, totally define the organization of lamellae within the radiating fibrils since this condition can be satisfied and still allow for rotation of a lamella about a radial axis of the spherulite.

Having outlined the basic nature and optical characteristics of polymer spherulites it is now possible to explain the origin of the *Maltese cross* extinction pattern displayed by spherulites in the polarizing microscope in general, and in Fig. 4.1 in particular.

It is well known, and the theory of optically anisotropic materials predicts, that in general any doubly refracting material rotated in its plane between crossed polars will appear to go dark or *extinguish* whenever light passing through it is vibrating along a 'principle axis' in the material. The directions of the principal axes depend on the atomic structure and are mutually at right angles to one another. Consider now a model or ideal spherulite composed of radiating fibrils for which the principal axes are along and at right angles to their length (Fig. 4.2). Imagine a thin diametric slice of such a model is examined in transmitted light between crossed polars. The fibrils positioned in the vibration direction for light entering the slice will be at extinction, as will those fibrils at right angles to the vibration direction. On the other hand all other fibrils will appear bright, particularly those in the '45° position' between the extinction directions. The net result is that spherulites in the slice appear to be doubly refracting but with an extinction cross, the arms of which are along and at right angles to the vibration direction of light from the polarizer.

In practice the structure of the spherulite will not correspond to the ideal and the slice or thin section will not in all cases be diametric. The appearance of the spherulite will then be modified and appropriate allowance must be made in image interpretation. It should also be noted that spherulites do not equate to the grain structure of metals. The lamellae are a closer approximation.

4.3. Characterizing polymer spherulites in the polarizing microscope

In this section it is assumed that the reader is familiar with the basics of polarized-light microscopy and the characteristics of optically anisotropic specimens.

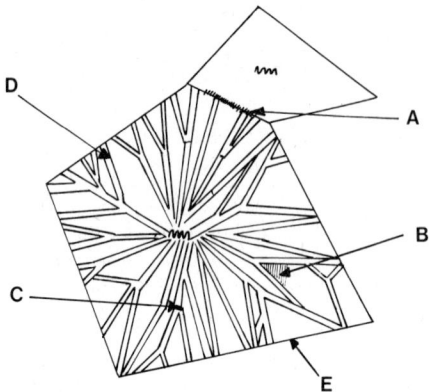

Fig. 4.2. Model of a polymer spherulite. (A) Interspherulitic amorphous regions; (B) Interfibrilar amorphous regions; (C) fibril defects; (D) fibrils; (E) polyhedral spherulite boundary (caused by the impingement of adjacent spherulites).

It is useful to be able to characterize the spherulites in a prepared specimen at least in subjective terms. Such characterization can help identify a polymer and provide information on the previous thermal history experienced by the polymer during processing. Some of the characteristics of the spherulites can readily be quantified; others are more conveniently left as subjective comparative assessments.

A number of characteristics of spherulitic texture are now considered. Imagine that a thin section of a plastics moulding is being examined. Points which might be commented on could include the following.

Spherulite size

Spherulites growing in bulk material are normally totally space-filling and, where they impinge, form straight or near-straight boundaries. In section these boundaries form a pattern of polyhedra. Clearly we can measure an average spherulite diameter or even the distribution of diameters. The mean spherulite size or inter-nucleus distance is related to the number of nucleating centres per unit volume which, in the case of homogeneous nucleation (see below), is in turn related to the crystallization temperature. In a moulding it is common to see small spherulites near the surfaces; larger ones deeper inside the material. Fig. 4.3 shows the typical gradation of spherulite size through the thickness of a moulded product.

Fibrilar texture

The spherulites of some polymers (e.g. polypropylene, Fig. 4.4) tend to show the fibrilar structure in the polarizing microscope more than others (e.g. polyethylene, Fig. 4.5). Although difficult to quantify, this characteristic is very useful in polymer identification.

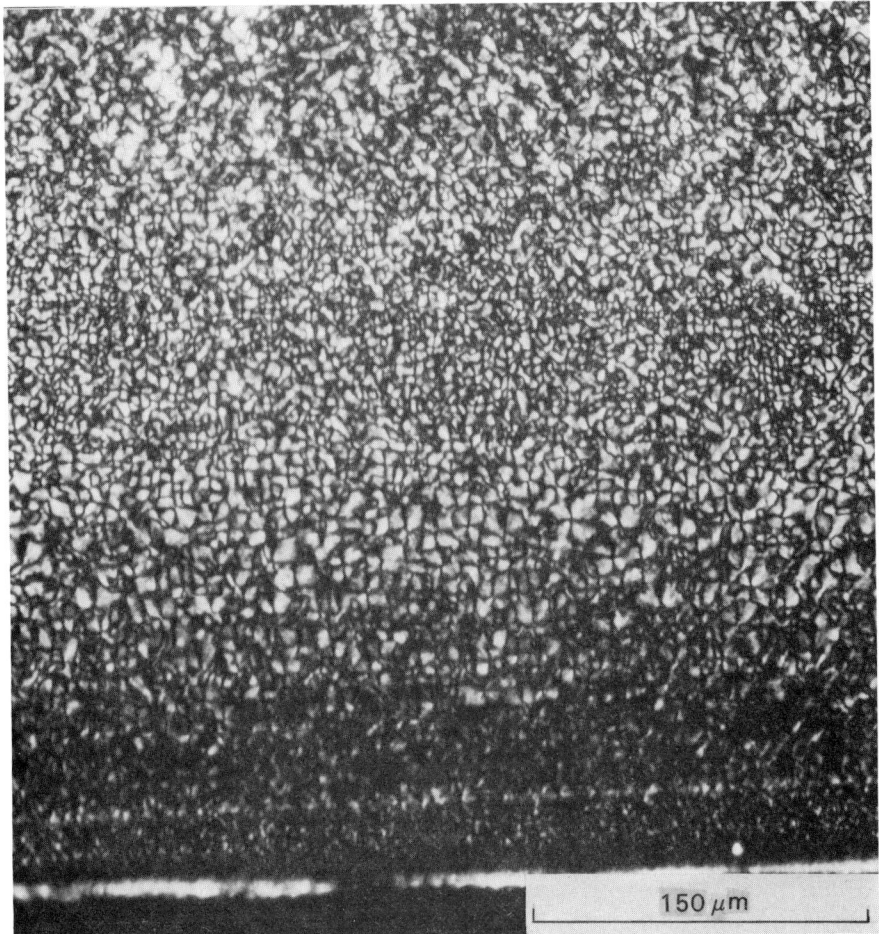

Fig. 4.3. Gradation of spherulite size in a polypropylene moulding (crossed polars).

Sign of birefringence

The birefringence of a spherulite is defined as $(n_r - n_t)$, where n_r and n_t are the radial and tangential refractive indices respectively. This quantity may be either positive or negative. Most semicrystalline polymers show spherulites all of one sign and this may then be used in identification. A few polymers display spherulites of either sign, often in the same section.

The sign of a spherulite is readily checked with a 1λ accessory plate. On inserting the plate into the microscope, opposite pairs of quadrants of the spherulite change colour. By consideration of the high-index direction γ of the plate and determining, by the resulting polarization colour, the directions of addition and subtraction, the high index direction, and thus the sign, of the spherulite may be deduced (see Fig. 4.6).

42 *Light microscopy of synthetic polymers*

Fig. 4.4. Spherulites in bulk polypropylene (crossed polars).

Magnitude of birefringence

By comparison with many minerals, or highly drawn polymer fibres, the birefringence $(n_r - n_t)$ of polymer spherulites is low. Even in thick sections or melt pressings (see Chapter 2) the polarization colour rarely exceeds the first-order yellow or orange. Commonly only low-order greys are displayed. When comparing spherulite birefringences, either quantitatively or qualitatively, it is better to use a thick section, say 40 μm in thickness, rather than a thin one since larger optical path differences are then produced. In thick sections the occurrence of higher-order colours help identify the material. For example, polyformaldehyde falls into this category, showing colours as high as second-order blue in thick section.

The measurement of spherulite birefringence presents many practical problems and generally only approximate results can be obtained. The main problem is that measurement methods depend on knowing the thickness of the feature being investigated. This is often indeterminate as it may not be known whether the

Observation of crystalline texture 43

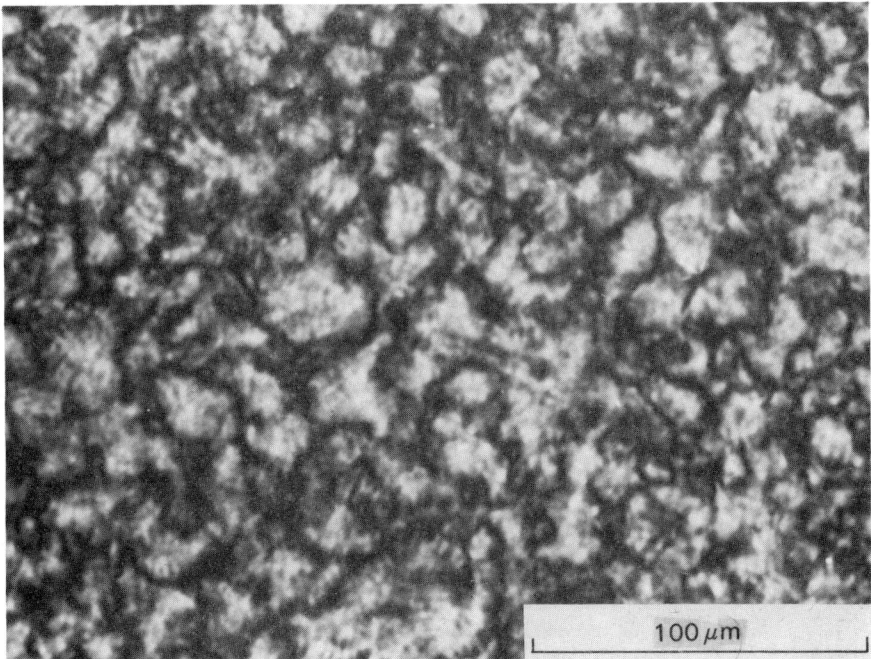

Fig. 4.5. Spherulites in bulk polyethylene (crossed polars).

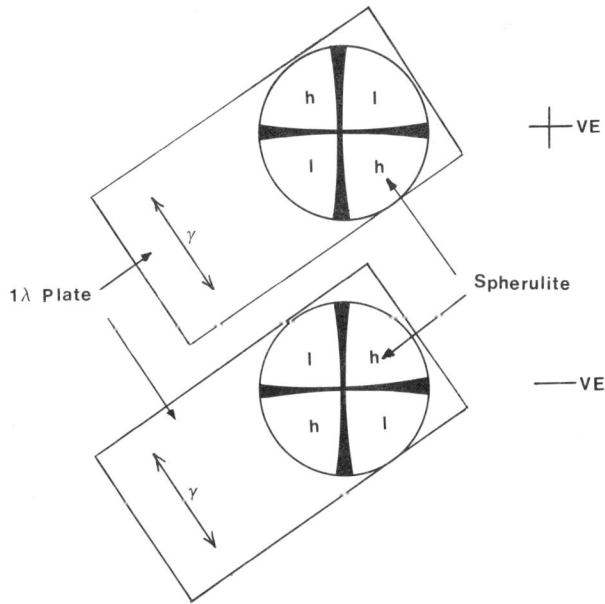

Fig. 4.6. Determination of the sign of a spherulite with a 1λ plate. (h) Quadrants with higher-order polarization colours; (l) quadrants with lower-order colours. Note that if the γ-direction of the plate is changed to be along its length, the quadrant colours are reversed.

Fig. 4.7. Fast-cooled 'granular' polyethylene (crossed polars).

feature extends to the full depth of the section or melt pressing, or whether the part of the spherulite being examined is a proper diametric section.

Spherulite perfection

Spherulites are often ill-formed and in some cases, such as fast-cooled polyethylene (Fig. 4.7), true spherulitic texture is not seen in the polarizing microscope; rather the material assumes an over all 'granular' appearance in polarised light. With the ideal spherulite model in mind the degree to which a specimen matches up to this model can be described in terms of the 'perfection' of the spherulites. In the absence of a ready method of quantifying this very vague characteristic it is inevitable that the use of the term 'perfection' will continue.

Ring morphology

Many polymer spherulites exhibit a ring structure, visible in the microscope as alternating dark and light rings (e.g. Fig. 4.8). Tilting experiments on such specimens suggest that, certainly in many cases, this occurs because of a co-ordinated twisting of the fibrils leading to a situation where periodically the observation direction is down an optic axis. This results in bands of zero birefringence. Although there is some dispute as to the precise nature and origin of this morphology, its occurrence is usually indicative of slow cooling of the material and as such is of value in investigating thermal history.

Observation of crystalline texture 45

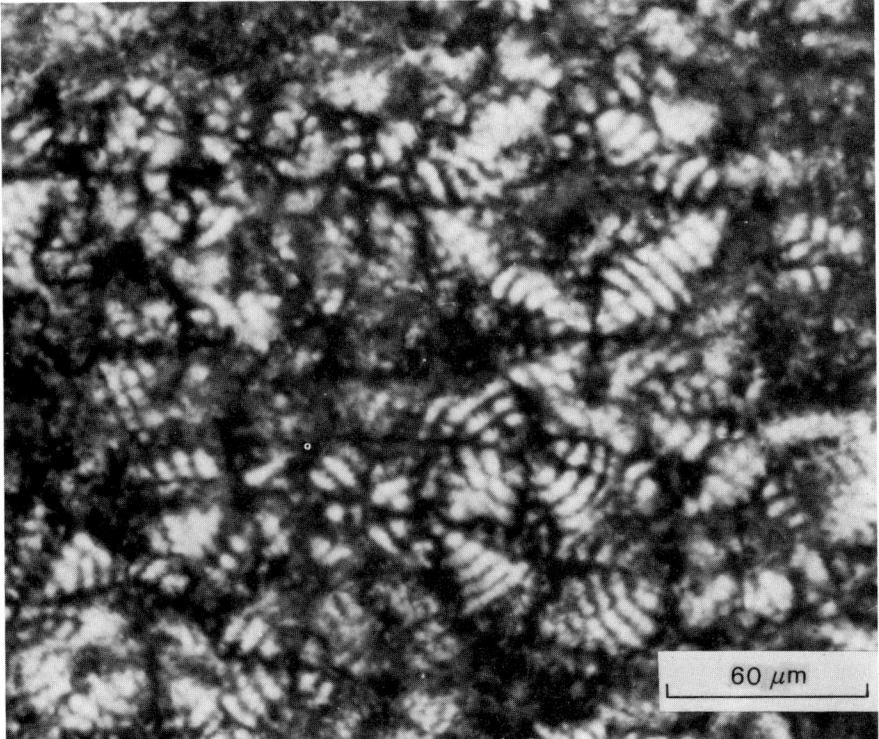

Fig. 4.8. Ringed polyethylene spherulites (crossed polars).

Extinction cross orientation

The above brief outline of the origin of the extinction cross shown by spherulites assumes that the principal directions in the fibrils are along and transverse to their length. There are some rare situations in which this does not apply and these can give rise to spherulites which have their extinction crosses inclined to the vibration directions of the polars.

4.4. A cautionary note

It is sometimes possible to be misled into believing that spherulitic structure exists when in fact no crystalline material is present. In practice most synthetic polymers will contain particulate additions or contamination. These can give rise to local radial stress distributions in the material which produce a birefringent patch showing an extinction cross. Clues that one is not seeing a true spherulite are that these effects are isolated, they show no fibrilar structure, and there is a steady gradation of birefringence at the edge of the feature and no sharp boundary (see Fig. 4.9.).

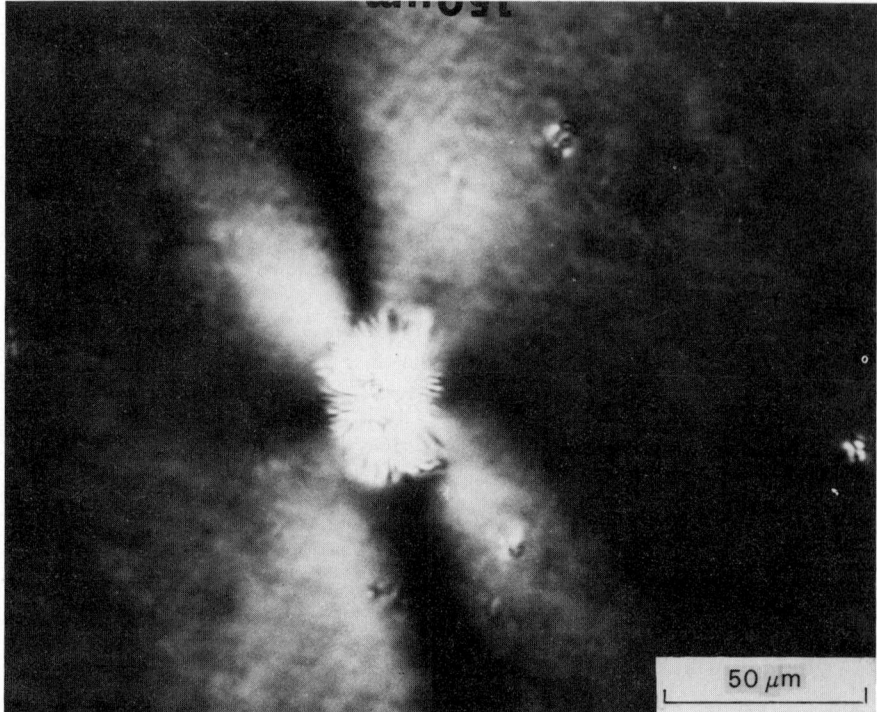

Fig. 4.9. False 'spherulite' in polymer film.

4.5. Alternative techniques of observation

As pointed out above, simple crossed polars are normally used for observing spherulitic texture. However, other optical methods of examination are possible, and may indeed make for easier acquisition of data.

Circularly polarized light

The incorporation into the microscope of a quarterwave ($\frac{1}{4}\lambda$) accessory plate with its slow or fast direction at 45° to the vibration direction of the polarizer allows the specimen to be observed in circularly polarized light. This plate is positioned between the polarizer and condenser. Some instruments have a slot for this, but in others the regular polarizer must be removed and replaced with a polarizing sheet and $\frac{1}{4}\lambda$ plate at a convenient point between the illuminated field iris and the condenser unit. A second $\frac{1}{4}\lambda$ plate is positioned between the specimen and the analyser (previously crossed with the polarizer). The orientation of the upper $\frac{1}{4}\lambda$ plate is such that an empty field of view should be as dark as possible. The effect of the upper plate is to convert the circularly polarized light back to

Observation of crystalline texture 47

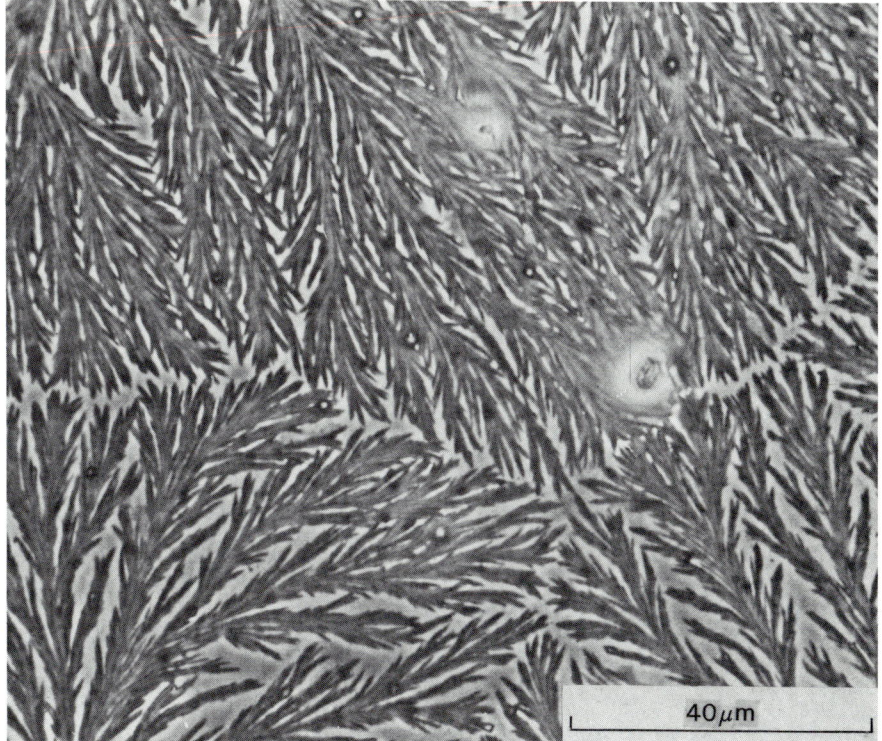

Fig. 4.10. Polypropylene spherulites (phase contrast).

plane-polarized which is then incapable of passing the analyser if the specimen is isotropic.

Spherulites viewed between circular polarizers do not show the characteristic extinction cross, although the other morphological features remain visible. This can have advantages when subjecting the image to automatic image analysis or when detecting optically isotropic features which might otherwise 'hide' in the extinction cross. Also it helps overcome the confusion which can sometimes arise between extinction crosses and spherulite boundaries.

Phase-contrast microscopy

Transmitted-light phase-contrast microscopy may be used to display spherulite morphology. For optimum contrast the preparations need to be thin ($\sim 5\,\mu m$) and especially free from cutting artefacts. Contrast is developed as a consequence of refractive index (= density) fluctuations in the specimen. Fibrilar texture is often particularly well displayed (Fig. 4.10).

Phase contrast can also be used to examine solution-grown single crystals of polymer otherwise almost invisible in the polarizing instrument.

48 *Light microscopy of synthetic polymers*

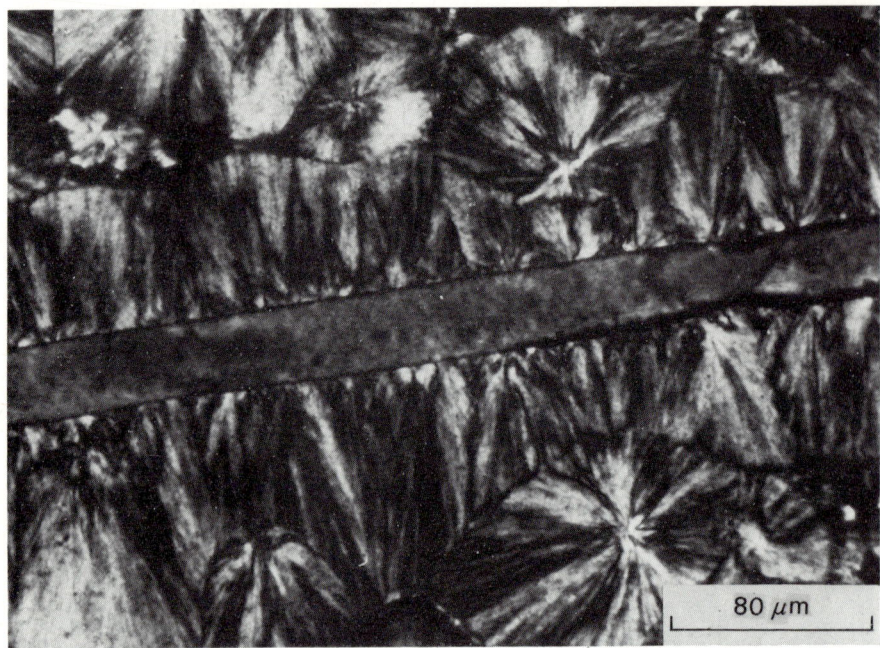

Fig. 4.11. Nucleation of polypropylene by a contamination fibre (crossed polars).

Transmitted-light differential-interference contrast

Although this technique will often appear to show spherulitic texture well, interpretation of the image can be difficult. For an isotropic specimen, the technique will normally display refractive-index fluctuations. In the case of spherulites, however, the situation is complicated by the optical anisotropy of these structures and the fact that the optical system is based on polarizing optics. The result is that, although contrast is developed, there is more than one mechanism at work and no unique one-to-one relationship exists between image contrast and microstructure. Caution in using the technique to analyse spherulite morphology is therefore recommended.

4.6. Nucleation effects

Up to this point it has been assumed that the crystalline spherulitic texture of bulk polymer results from nucleation of the material by itself. This homogeneous process is a statistical one strongly dependent on the temperature of the melted polymer and naturally gives rise to a distribution of spherulite sizes.

On the other hand, it is possible, indeed quite common, for heterogeneous nucleation to take place as the result of contamination, the presence of additives, or contact with a nucleating surface.

An example of nucleation by contamination is shown in Fig. 4.11. Obviously the

Observation of crystalline texture 49

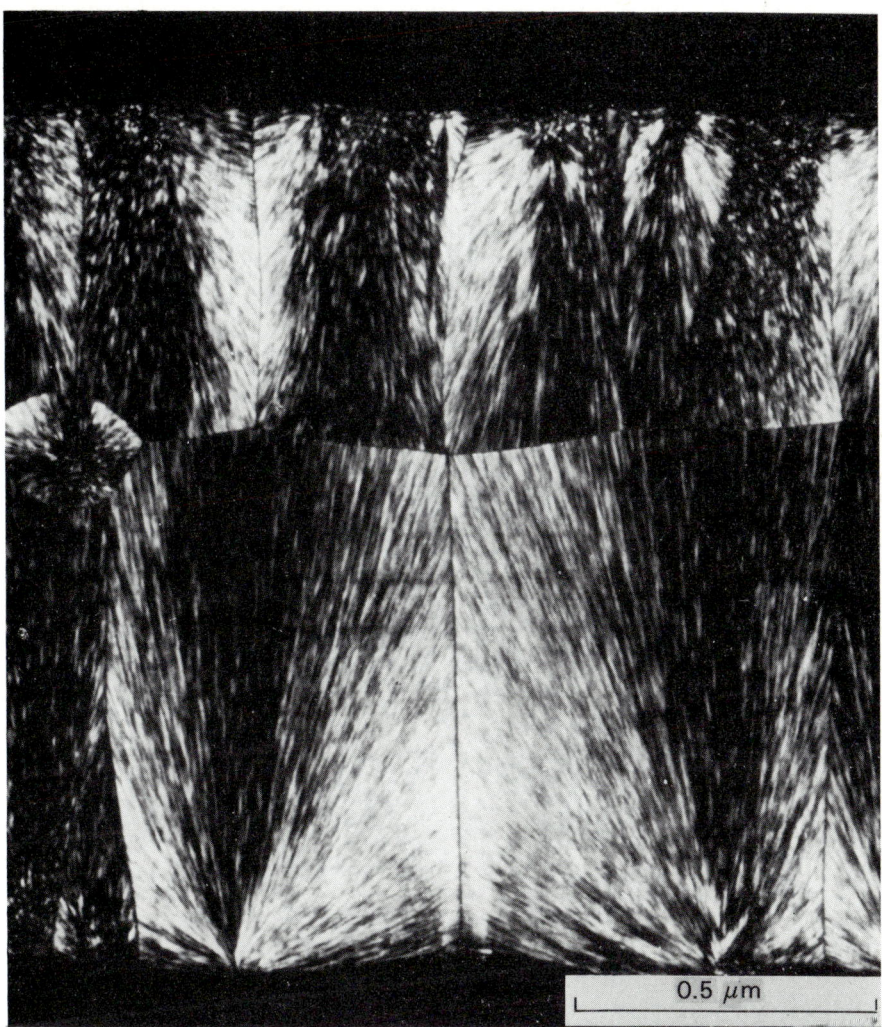

Fig. 4.12. Surface nucleation of polypropylene (crossed polars).

effect of such nucleation is local and readily recognized. A similar situation can occur with surface nucleation, which is of course again localized. This effect can be extreme (Fig. 4.12) and in such cases can lead to substantial weakening of the product.

Nucleation by additives may be intentional or unintentional. Assuming the particulate additive is well dispersed, the result is a small and even spherulite size throughout the product, except at very-fast-cooled surfaces. 'Nucleated' grades of nylon and polypropylene are commercially available as they offer certain processing advantages.

Unintentional nucleation may arise from pigments and, in instances where it is

50 *Light microscopy of synthetic polymers*

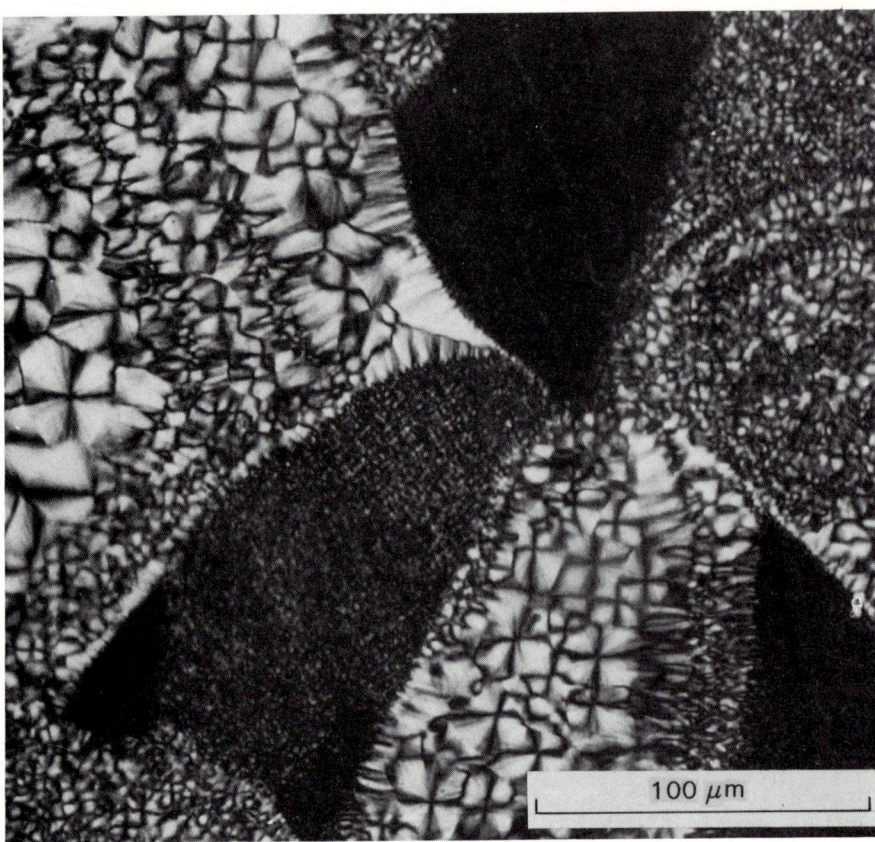

Fig. 4.13. Uneven mixing-in of pigment into polyethylene giving rise to patchy nucleation (crossed polars).

poorly dispersed, the effect on crystalline texture is most noticeable. *Low shear* processing such as compression moulding or rotational casting in which there is little mixing of blended pigments tends to illustrate the effect well (Fig. 4.13) and products may suffer adverse mechanical properties as a result.

Another nucleation phenomenon is that of *row nucleation* arising from crystallization from a melt in which there is aligned rather than random molecular orientation. This and similar situations give rise to crystalline regions containing extended molecular chains and these in turn act as linear nucleation sites for more normal spherulitic crystallization. In the light microscope we see growth from a row of nucleii.

4.7. Hot-stage microscopy

A micro hot-stage fitted to a polarizing microscope significantly extends the usefulness of the instrument for work on crystalline polymers. The main applications follow.

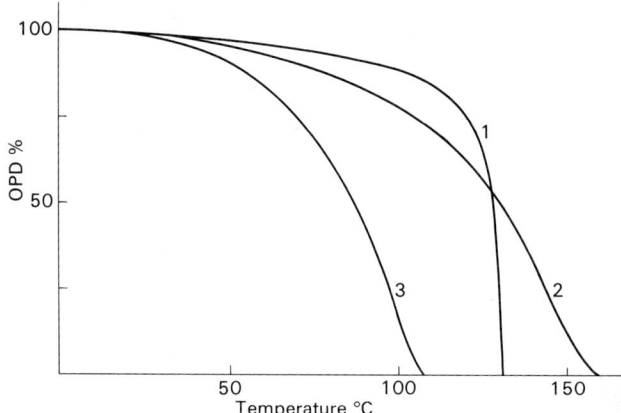

Fig. 4.14. OPD/Temperature curves for (1) H.D polyethylene; (2) polypropylene, and (3) L.D polyethylene.

Determination of melting points

An 'optical melting point' can be defined as the temperature at which, on heating the specimen, the last traces of spherulite birefringence disappear. In practice the spherulite birefringence will be seen to fall over a range of temperature prior to the final melting point. The width of this range, as well of course as the optical melting point, will vary from one polymer to the next and is a powerful identification aid. Although it is quite feasible to do this work subjectively, the precision and accuracy are greatly improved if the eye is replaced by a photometer. Graphs relating spherulite optical path difference (i.e. thickness × birefringence) can be produced which are very characteristic of a given polymer (Fig. 4.14).

The heating rate used is important on two counts. First, excessively high rates may mean that the specimen lags behind the indicated temperature of the stage. This is a function of stage design and specimen size. Second, the polymer itself will take a finite time to come to equilibrium at a given temperature. As a result heating rates around 1 to 3°C/min. are commonly used although as high as 10°C/min may be employed if lower accuracy is allowable and for 'sighting shots' prior to a slower run.

Selected optical melting points for some common semicrystalline polymers are given in Table 4.1. Note however that some variation will occur between those from different manufacturers or made by different processes.

Study of crystallisation behaviour

The growth of spherulites and the rate of occurrence of nucleii can also be studied with a hot stage. This can provide information of value to processors. Normally the

Table 4.1. The optical melting points of some common polymers

Polymer	Melting point (°C)
Polyethylene high-density	135
Polyethylene low-density	118
Polypropylene	168
Nylon 6	220
Nylon 11	185
Nylon 6:6	260
Poly acetal	180
Poly (4 methyl pentene 1)	245
Poly (ethylene terepthlalate)	270
Poly(tetrafluoroethylene)	330

experiment will be an isothermal one in which a melt-pressed specimen is rapidly transferred from a hotplate, set well above the melting temperature of the polymer, into the hot stage set at a known crystallization temperature below the melting point. The number and diameter of spherulites is a constant for a given temperature, at least until the late stages of crystallization. A series of such experiments can be used to build up a crystallization rate vs. temperature curve for the polymer. In practice the rate of crystallization may be very fast and a large number of nucleii are involved, even at temperatures only a little below the melting point. High-speed recording systems (e.g. video tape, cine film) may help to resolve the problem, but even then the method may become impracticable for other reasons such as the excessive superimposition of features in the image.

Detection of phases in blends and copolymers

Blends of semicrystalline polymers and certain 'block' copolymers often exhibit phase separation on a scale within the resolution range of a light microscope. Detection of phases may be aided by heating the material in the hot stage. In a two-phase system, for example, a temperature can be reached when one phase is melted and the other still crystalline and birefringent. At this point the phase separation becomes very much more evident and easier to observe. Identification of both phases is aided by determining their melting points.

The design of a micro hot stage is necessarily a compromise between the need for good thermal stability and accurate specimen temperature determination on one hand, and optical accessibility on the other. In particular, difficulties can arise if the designed objective working distance is large. Most hot stages will work with a standard 10× lens or lower. Higher resolution involves the use of long-working-distance objectives and condenser units. However, these often have numerical apertures little larger than lower-power lenses and their use may not provide as much resolution advantage as might have been expected or hoped for.

Another problem is that, with some polymers melting at temperatures as high as 300°C, the objective and condenser lenses may well become excessively warmed. Repeated temperature cycling may affect the lens mounts and introduce strain into the lens. This then degrades the performance of the lens for crossed-polars work as

Observation of crystalline texture 53

the glass is now birefringent. Again the use of long-working-distance optics mitigates the problem, as does forced air cooling of the lens.

4.8. An optical diffraction method for the study of spherulites

The size of spherulites in a field of view may be determined by using a diffraction technique. The method is of interest because of a number of useful features.

1. An average spherulite diameter for the whole field is calculated.
2. The size range measurable extends from 0.05 to 10 μm, i.e. it goes below the normal limit of resolution.
3. Comparatively thick specimens (up to about 25 μm) can be investigaged, either as melt pressings between slide and coverslip, or as thin sections. In the latter case particular care must be taken not to strain the specimen.

The technique involves using a standard polarizing microscope with crossed polars. The specimen is mounted on the stage and focused in the normal manner; then both the illuminated field and aperture irises are almost fully closed. The back focal plane of the objective, usually a × 40 high-numerical-aperture lens, is observed

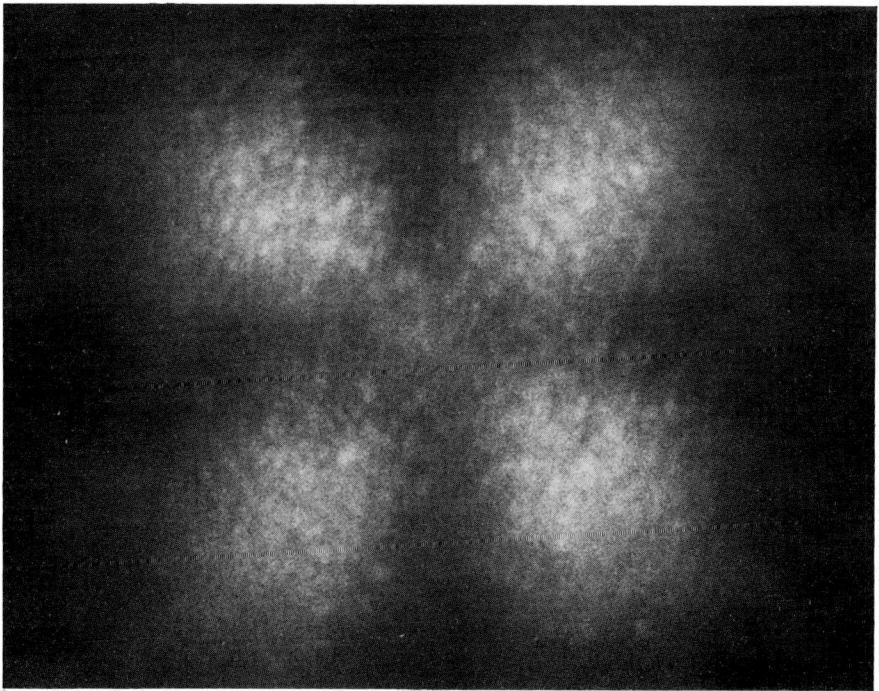

Fig. 4.15. Back focal-plane light-scattering pattern for polyethylene (crossed polars).

using either the Bertrand lens if fitted to the instrument, or a telescope such as is provided for the adjustment of phase-contrast instruments. The former is to be preferred since it is an advantage to have an eyepiece measuring graticule in focus at the same time as observing the back focal plane, and these are not normally available in telescopes.

A crystalline polymer showing good spherulitic organization will show a 'clover leaf' intensity pattern in the back focal plane, such as shown in Fig. 4.15.

Small spherulites scatter at high angles to the unscattered beam direction so that the diameter of the crossed-polars diffraction pattern increases as spherulite size reduces. This inverse relationship is expressed by

$$\bar{R} = \frac{K}{\sin \theta/2}$$

where $\bar{R}$ is the mean diameter of spherulites in the field of view, θ is the angle at which maximum scattering intensity occurs, and K is a constant dependent on instrumental parameters, the wavelength of light used, and the optical properties of the polymer.

Since the scattering pattern is of low relative intensity, it is prudent to use a high-intensity light source especially if photography is contemplated. It is also advisable to use a monochromatic filter.

It should be noted that this technique is not the same as conventional conoscopy in that the specimen is illuminated with a thin pencil of parallel rays, not a highly divergent beam. The necessary requirement for a high-numerical-aperture objective is the same however. The technique can be used in conjunction with a hot-stage to follow spherulite growth rates.

5

Blends, composites, and copolymers

5.1. Introduction

Many plastics and rubber articles are today not manufactured from a *pure* polymer but from a mixture of polymer with polymer or polymer with an inorganic material or *filler*. If in the case of polymer–polymer systems the mixing is at the molecular level, and thus below the resolution limits of the microscope, we will define these materials as *blends*. Mixing on a coarser scale means that there are local areas of homogeneous compostion, or *phases*, which can be identified and characterized in the microscope. Such materials will be termed *composites* and are of particular interest to the microscopist. The nature and distribution of the phases is of prime importance in determining the physical properties of the composite and thus its usefulness as a commercial material. Composite plastics conventionally also include the mixtures of polymer with inorganic mateials, such as polymers containing glass fibres or other mineral fillers.

Since the methods of preparation and examination of polymer–polymer composites are generally very different from those for polymer–inorganic composites, these are treated separately below. Polymer–polymer systems are discussed as *soft* composites and the others as *hard* composites.

Copolymers are materials in which the mixing in effect takes place within the polymer chain. By polymerizing with more than one monomer present the sequence of monomer units, say A and B, in the chain can be controlled and varied. For example the occurrence of A and B in the chain may be random. In this case microscopically there is nothing to distinguish the material as a copolymer on a phase separation basis. However if the sequence contains long runs of A or B, then the material may show local variations in composition visible in the microscope. In this situation the plastic or rubber shows many of the microscopical characteristics of a composite.

Propylene–ethylene copolymers present a good example. If the ethylene units are distributed randomly in what would otherwise be a polypropylene chain, a material is obtained which exhibits a single phase (ignoring the fact that both crystalline and amorphous regions will be present). On the other hand, long runs of ethylene units in the chain can, according to how the material is processed, give rise to 'droplets' or a separate phase of ethylene-rich material in a propylene-rich matrix. This ethylene-rich material may, in rare instances, show polyethylene-type spherulitic texture.

For the purpose of discussing blends, composites, and copolymers in detail it is therefore convenient to divide these materials up into

Blends and random copolymers;
Soft composites and block copolymers — capable of being thin-sections or melt-pressed;
Hard composites — generally needing metallographic or petrographic techniques.

5.2. Blends and random copolymers

As indicated above, these materials show no phase separation. Light microscopy is therefore limited to comment on any crystalline texture or molecular orientation effects observed with a polarizing microscope, and refractive-index determinations.

Finding the refractive index by means of the Becke Line Test and a range of suitable immersion liquids can assist in establishing the composition of a blend. A common light microscope working with a × 20 or × 10 objective is used. For a two-component system in which both blended polymers are amorphous the refractive index of the blend will be a linear function of composition, i.e.

$$(v_1 + v_2) n_b = v_1 n_1 + v_2 n_2$$

where v_1 and v_2 are the volumes of the two polymers blended and n_1 and n_2 are their refractive indices. The refractive index of the blend is n_b.

In the case of blends of polymers in which one or both of the constituent polymers normally crystallizes, a more complicated pattern of refractive-index behaviour occurs. The introduction of the second material into the first may initially destroy crystallinity and produce a non-linear change in refractive index. This is illustrated by Fig. 5.1 which shows the refractive-index behaviour of a poly(vinyl chloride) (PVC)-poly(methyl methacrylate) (PMMA) melt blend. At low concentrations of PMMA a nonlinear relationship between concentration and refractive index occurs as the small amount of PVC crystallinity (about 10 per cent) is destroyed. Once this has happened a linear relationship is established for higher PMMA concentrations. Similar effects are seen in the case of random copolymers.

5.3. Soft composites

The examination of thin sections of soft composites is usually one of the more rewarding branches of polymer microscopy. The information about specimens which can be obtained and which may be of direct relevance to the performance of the polymer in service includes

1. The size and shape of phases;
2. The volume distribution of phases throughout the specimen and the nature of their boundaries;
3. The composition of the phases (usually via refractive index measurement or melting points);

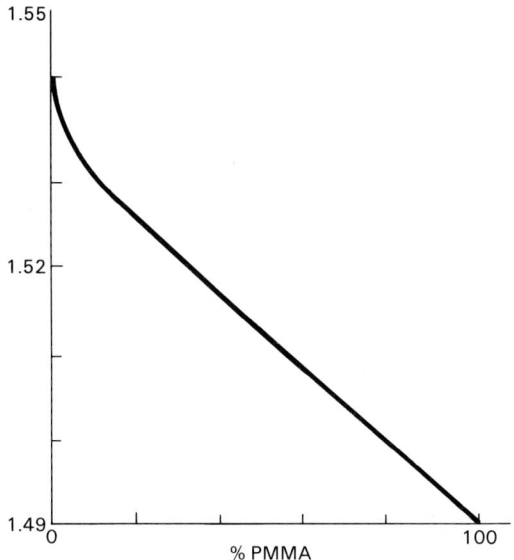

Fig. 5.1. Refractive index (RI) vs. percentage poly(methyl methacrylate) (PMMA) content.

4. The presence of any phase inversion;
5. Any preferred phase orientation;
6. Crystalline texture or molecular orientation in phases;
7. Correlation between polymer phases and additive (e.g. pigment) distribution.

The problems associated with the microscopy of 'soft' composites are in many respects similar to those encountered by biologists working with tissue. Since plastics generally do not accept stains, contrast can be developed only by optical methods, exploiting the differences in refractive indices between phases. Often these differences are small and require sensitive systems. These include

1. Dark field;
2. Negative or positive phase contrast;
3. Differential interference contrast;
4. Two-beam interference methods.

A section of a typical soft composite imaged using each of these transmitted light modes is shown in Fig. 5.2. If one of the phases is semicrystalline and exhibits crystalline texture then clearly polarized light can be added to this list.

Special problems in the use of these techniques on polymers may be encountered, some related to specimen preparation and others to the composites themselves.

1. *Sectioning problems.* The procedure for sectioning polymers was discussed in Chapter 2. Good section quality, particularly in respect of freedom from knife marks, is required for phase contrast and interference methods. The optical

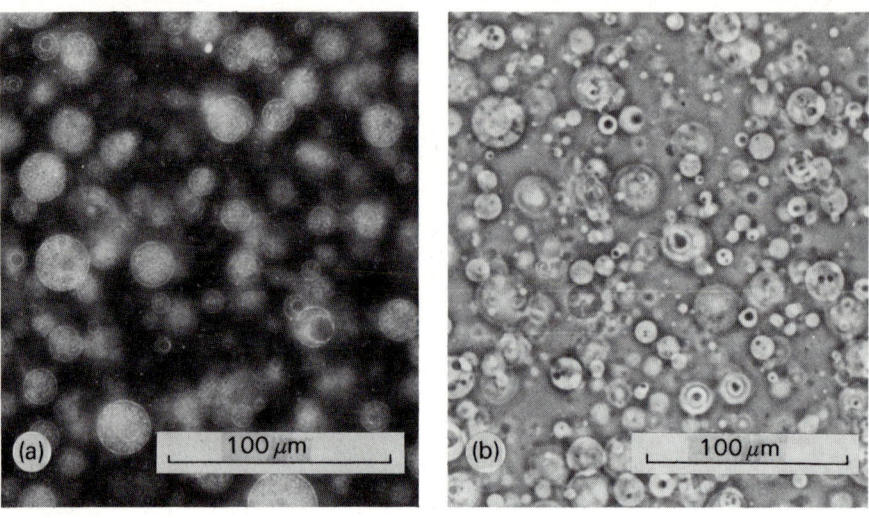
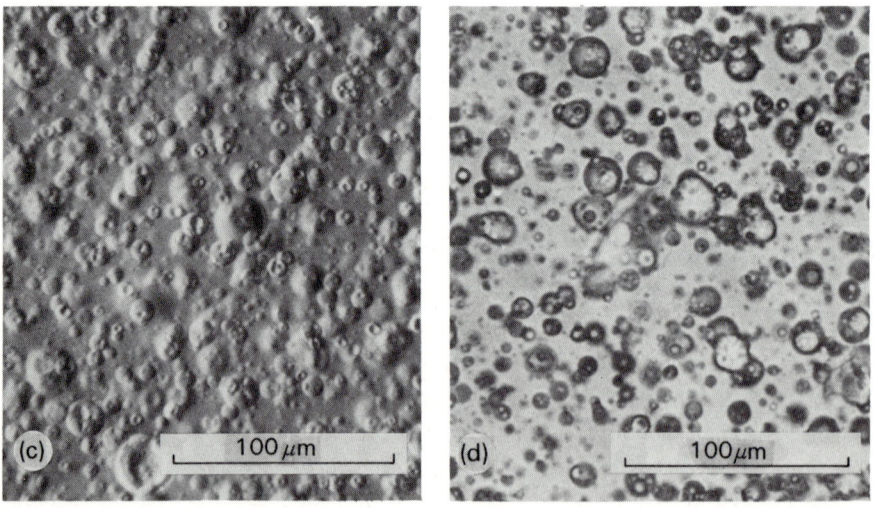

Fig. 5.2. A polymer–polymer composite imaged in transmitted Light by (A) dark field; (B) phase contrast; (C) DIC; (D) Jamin–Lebedeff two-beam interference method.

path-length variations in the section arising from refractive-index differences, which are the basis for contrast development, may be swamped by variations in the geometrical thickness of the section due to a poor knife edge. Normally this effect would be minimized by mounting the section in a liquid whose refractive index matches that of the dominant or continuous phase. In practice the danger of chemical interaction between polymer and liquid (swelling, relaxation, solvation, etc.) may adversely restrict the liquids available and some index mismatch must be tolerated.

Another sectioning problem which is encountered is that any microvoiding or *stress whitening* of the polymer is especially visible using the above techniques on account of the comparatively large optical path differences which may result. Again the true microstructure in the section may be masked.

Both problems are overcome, or at least reduced to acceptable levels by close attention to section quality.

2. *Optical problems.* When examining a composite for the first time one may have no knowledge of the magnitude of the optical path-length variations in the specimen. These will depend on both the size and relative refractive indices of the phases. As an example, consider a $10\,\mu m$ thick section of a composite consisting of a poly(methyl methacrylate) matrix (refractive index $n = 1.495$) containing 5 per cent of polybutadiene ($n = 1.520$). A spherical droplet of diameter $2.5\,\mu m$ shows a maximum optical-path difference of $(1.520 - 1.495)\,2.5 = 0.00625$ microns. Expressed as a fraction of a wavelength of green light ($= 0.5\,\mu m$) the optical path difference is $0.0625/0.5 = 0.125$, i.e. $\lambda/8$. Any of the above techniques will readily generate contrast from such a droplet. Significantly smaller droplets or a smaller refractive index difference would favour the use of phase contrast rather than differential interference contrast, because of the high sensitivity needed. Much larger droplets could in fact *reduce* the contrast developed using phase-contrast methods and differential-interference contrast methods would be more applicable. The clear message is that it may not be possible to judge, when confronted with an unknown composite, which method of contrast enhancement is going to be the most successful.

In the situation when the droplet size is large relative to the section thickness, it may be assumed that the dispersed or droplet phase extends through the total thickness of the section. There is then control of the optical-path differences exhibited by varying section thickness. Given a phase-contrasting system showing certain optical characteristics it should be possible to optimize the contrast. It is not uncommon to find an increase in contrast in composites being attained by *reducing* section thickness.

Like phase contrast methods, dark field systems are sensitive to all but the smallest refractive-index differences (say 0.0001) but the latter suffers the disadvantages of low image light intensity and contrast development only at the perimeter of phases.

The two-beam interference methods are used essentially for quantitative work, particularly measurement of refractive indices for phase identification. Discussion of how this is done is outside the scope of this text, but the use of such methods for contrasting phases poses certain problems. The generation of high interference contrast in these instruments normally involves restricting the condenser aperture by slit or stop, to the detriment of resolution. This may be an undesirable and unnecessary trade-off for qualitative work, or in quantitative studies involving only the spatial distribution measurement of phases.

3. *Interpretational problems.* In the study of polymer composites one has to be constantly aware of a number of factors which bear upon the interpretation of the image and the conclusions drawn. Some of these problems concern sampling. 'Sorting' effects during extrusion or moulding can easily produce gradations in both the size and shape of the droplet or discontinuous phase in a composite. Great care is then necessary to ensure the statistical significance of droplet concentration and geometrical parameter measurements. Other problems are stereological in nature. Fig. 5.3 shows an extruded composite in which there has been apparent elongation of the droplets. The section has been cut in the extrusion direction. Only a second section cut at right angles to this one will reveal whether these distorted droplets are now rod-like or plate-like. The difference could be very significant in relation to the performance of the extruded product.

Yet further interpretational problems arise if the concentration of the droplet phase is high; say greater than 10 per cent by volume. There is then considerable overlap of information if the droplet size is small and observations on individual droplets will be difficult. The reported small depth of field of the differential-interference contrast methods may then become a distinct advantage, allowing optical sectioning to different layers of droplets. In the absence of a good microtome, one may be forced into examining melt-pressed specimens as discussed in Chapter 2. It is unlikely that such preparations will be sufficiently thin for good phase-contrast work, but differential-interference methods are more successful because of the optical sectioning effect.

Listing the problems of studying polymer composites might suggest that these are particularly difficult materials. In practice care with specimen preparation and some thought on the selection of technique can yield valuable morphological information. For example the enhanced mechanical properties of rubber modified composites rely on the droplet phase being neither too small nor too large (the normal average size range is 1 to 5 μm), there being a random distribution and no undue deformation of the droplets. Many departures from the ideal are observed!

5.4. Block copolymers

It has already been mentioned that copolymerization involves the polymerization of two monomers and that the result can be molecules having long runs or 'blocks' of identical chemical repeat units. These block copolymers (or *terpolymers* if three monomers are involved) may exhibit microstructure on scales of interest to both light and electron microscopists. Just how blocks can organize themselves to display a microstructure similar to composites is a subject of much research. Certainly

Blends, composites, and copolymers 61

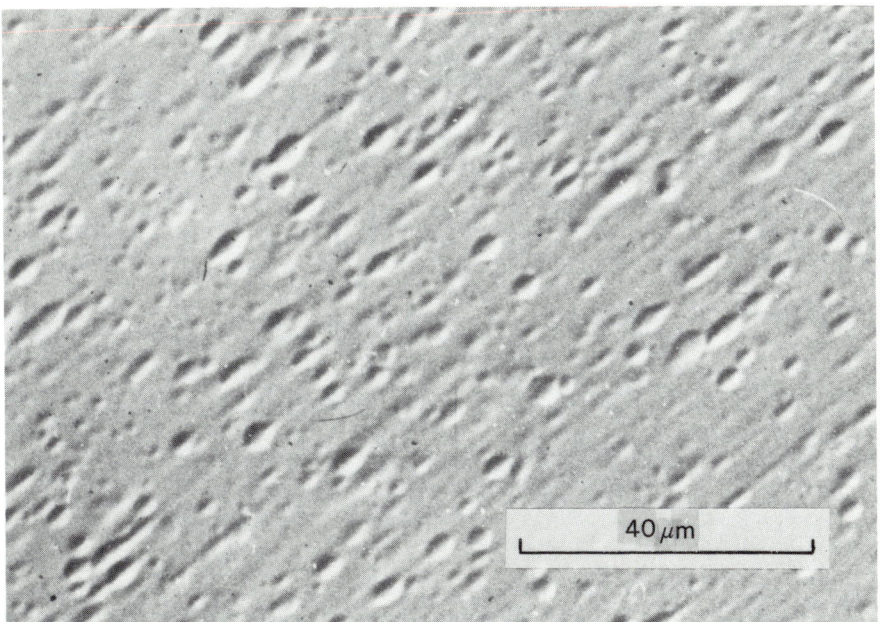

Fig. 5.3. Spatially oriented droplets in a polymer–polymer composite (transmitted-light DIC).

the occurrence of a droplet structure in a block copolymer is often strongly related to processing history (more strictly the thermal and melt shear stress history).

At the electron-microscope level, the phase-separated regions are more commonly and correctly referred to as 'domains' rather than droplets. Indeed, geometrically regular arrangements such as plates and rods are common in certain materials.

In the light microscope, phase-separated block copolymers may be treated and examined in much the same way as soft composites, but since the dispersed phase size is usually smaller than in true composites, higher-resolution work is involved.

It is possible to employ a light microscope (using crossed polars) to indirectly study the domain structure of copolymers at what would normally be regarded as the electron microscope level. The domain microstructure is similar to that found in biological (particularly botanical) specimens and which has been studied via the *form birefringence* that such materials can exhibit.

Viewed between crossed polars, some block copolymers (styrene–butadiene–styrene is a good example) also show form birefringence arising from the organization of the domains into arrays of plates or rods in a continuous matrix. The rather diffuse bright patches showing in Fig. 5.4 are areas of such organization. Which form of structure is present depends on the styrene–butadiene ratio and, again, on the processing history.

By determining whether the birefringent areas are negative uniaxial or positive uniaxial in character the material is shown to consist of plates or rods, respectively. In the case of plates the optic axis is perpendicular to the plane of the plates. For

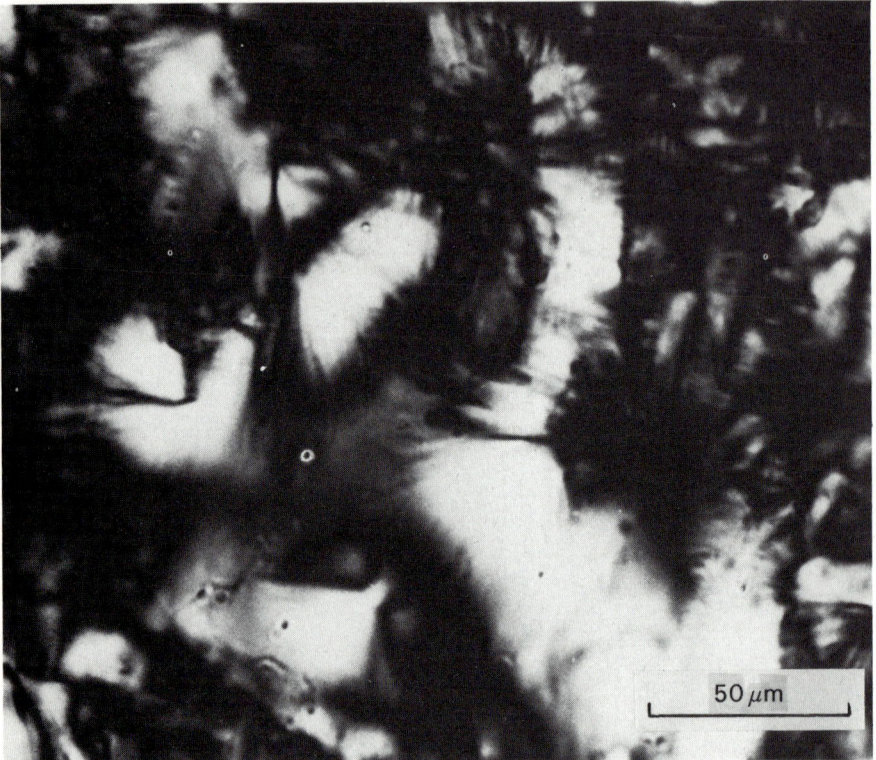

Fig. 5.4. Form birefringence in a styrene–butadiene–styrene copolymer (crossed polars).

rods the optic axis is along the axis of the rods. Such an examination can be of considerable use as a precursor to an electron microscopy study, or as a parallel, confirmatory, observation.

5.5. Hard composites

Hard composites include glass-filled polymers. The glass may be present as a mat as in glass-reinforced polyester (GRP) or as chopped strand glass 1 to 10 mm in length. In the latter case the inclusion of the glass in the polymer, even at the 20 or 30 per cent level, does not prevent it being extruded, injection-moulded, or otherwise processed using standard techniques. Nylon, polypropylene, polyacetals, and polycarbonate are typical materials available as short-fibre composites. In the case of GRP and other composites of similar construction, the position of the glass is fixed during production by a lamination and impregnation process. In short-fibre melt-processed materials, the fibre distribution and orientation is a function of the processing conditions, the viscosity of the polymer, and the concentration. The spatial orientation of glass is an important factor in determining the mechanical properties of such composites and is therefore examined microscopically.

Sheet-moulding compounds (SMC) are manufactured and moulded with the glass fibres in a two-dimensional, usually random distribution. Unidirectional layers are employed for additional strength but are used with random chopped-strand layers.

Fibre lengths less than 20 mm are not used and lengths in excess of 50 mm do not improve properties significantly. Mixtures of continuous filament and chopped strand are now used to reduce fibre orientation or to enhance properties in a particular direction.

Our definition of hard composites also embraces polymers with high levels of mineral filler such as talc, calcite, or graphite. At low additive concentrations, some polymers can be successfully sectioned and treated as *soft* but over about 20 per cent concentration the additive cause both sectioning and observational difficulties. Indeed it should perhaps be mentioned that even short glass-fibre-filled polymers can be thin-sectioned under certain circumstances. If the concentration of glass is relatively low, if the polymer itself sections readily, and if some disturbance of the polymer by the additive during sectioning can be tolerated, then this technique may be applicable. Such preparations can be examined using transmitted light. However the more common approach to the examination of hard composites is to use reflected-light techniques.

5.6. Fibre composites

For preparations containing glass fibre, sufficient reflectivity difference normally exists between polymer and glass for adequate contrast to be obtained in common light. Contrast may be marginally improved if the diameter of the field iris (*not* the aperture iris) is restricted to below its normal setting, the penalty being a restricted illuminated field. The reflected light, dark-field method may be applicable if only the glass–polymer interface is of interest. In cases where the polymer has been etched or dissolved away to leave the glass proud of the polymer surface, the use of differential interference contrast is advantageous for contrast generation but may confuse quantitative analysis of the material.

The glass fibre used in polymer reinforcement is very commonly circular in cross-section and of fairly uniform diameter. This considerably simplifies the quantitative analysis of the images of polished surfaces. The exposed cross-section of glass fibres will in general be an ellipse. The angle α the major axis makes with a chosen reference direction, such as the edge of the sample, is one parameter necessary to describe orientation. A little thought will confirm that two angles are needed to describe the orientation of the glass and the second is derived from the ellipticity of the exposed cross-section. It can readily be shown that the angle of tilt of the fibres β referred to a line at right angles to the plane of the polished surface is given by

$$\beta = \sec^{-1}\left(\frac{L}{D}\right)$$

where L and D are the major and minor axes, respectively, of the exposed ellipse. Measurement, for a large number of fibres, of the ratio L/D and the angle α will therefore yield fibre-orientation information. There are however a number of inherent problems in this method. First, the method does not distinguish fibre inclined equally either side of the normal (perpendicular) to the plane of the cross-section, i.e. two possible inclinations will correspond to a given result for β. The practical significance of this ambiguity can be minimized by the choice of polishing plane. For example, if the plane of observation is at right angles to the direction of strain experienced by the composite in service or during testing, then the contribution to the properties of the composite by fibres in either of the two ambiguous orientations will be the same, so that we do not need to distinguish between them. Alternatively, orientation information can be collected for two observation planes, usually at right angles to one another. In this case a full description of the analysis of the results is outside the scope of this text and it is recommended that a reader faced with the problem consults a good text on stereology.

The second problem arises because the probability of intersecting a given fibre with a given observation plane will depend on the fibres orientation, being a maximum when the fibre is at right angles to the plane and a minimum when it is parallel to it. The data acquired is therefore biased towards the former and a correction is necessary to the apparent orientation distribution if the true distribution is to be obtained. Again a text on stereology is recommended before quantitative work is undertaken. However even when carrying out only a qualitative image interpretation this distortion should be allowed for.

The fibres in a long-fibre composite can be considered as being of infinite length from the standpoint of image interpretation. This is not the case with a short-fibre composite and this forms the basis of the third problem. A chopped fibre viewed perpendicular to its axis will appear rectangular not elliptical because of its finite length. Again the practical method of mitigating this complication is to choose a plane of observation which minimizes the occurrence of such fibres.

Presentation of the numerical results of orientation distribution assessments can be done in a number of ways. Perhaps the most convenient is to use the approach shown in Fig. 5.5 based on the use of readily available polar co-ordinate graph paper.

In the case of composites involving the use of fibre woven mats, the glass orientation will obviously vary in a regular manner within the specimen. Injection-moulded or extruded short-fibre composites may show a considerable variation in fibre orientation through the thickness of a manufactured product and from place to place within it (Fig. 5.6). To fully characterize such a product is therefore a major task involving great attention to sampling and interpretation of results. It might be added that alternative methods of quantitative analysis of these materials, such as microradiography or scanning electron microscopy of etched surfaces, offer little advantage at the analysis stage.

Although a measurement of total glass content can be made from the exposed area of glass in a polished section (again, certain stereological corrections are

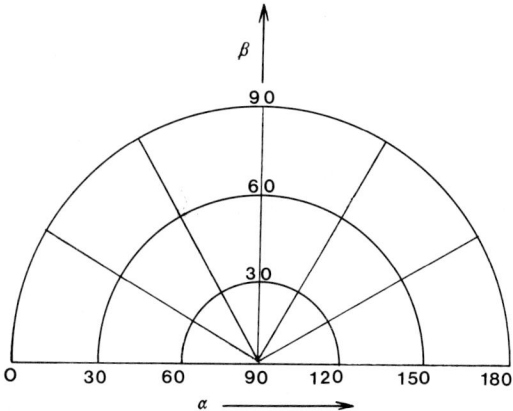

Fig. 5.5. A method of plotting fibre-orientation results.

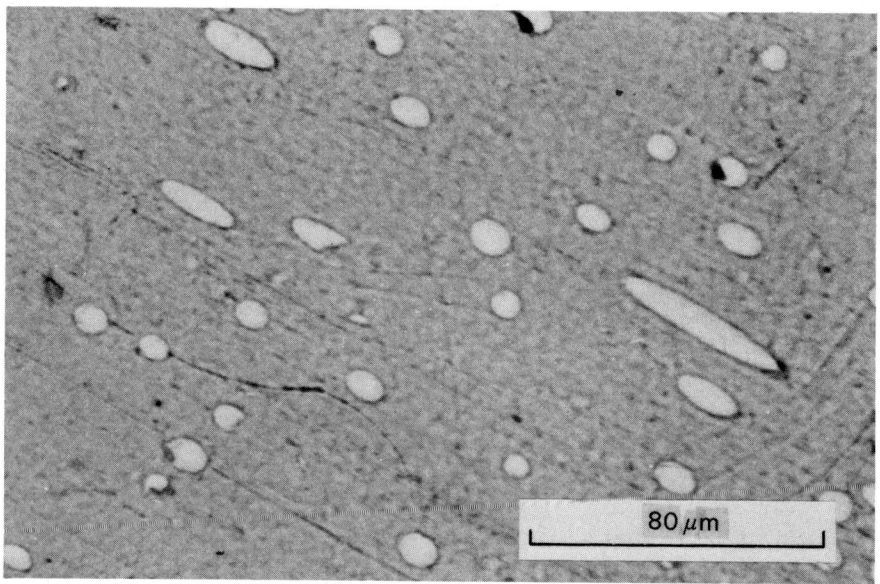

Fig. 5.6. Random glass-fibre orientation in a short-fibre injection-moulded polypropylene (reflected light).

necessary) this is better carried out non-microscopically by solvation or pyrolization of the composite and by weighing.

Assessments of glass-fibre length distributions are also aided by first removing the polymer by one of the methods mentioned above, then applying normal micromeasurement techniques to the residue. In practice the distribution of glass length is not very wide, even after processing the composite, so that statistically significant results are often obtained without the need to measure excessively large numbers of fibres. For measurement, and in order to reduce confusing refraction effects, the glass residue is normally mounted in a fluid of similar, but not matching refractive index before being examined by transmitted light.

5.7. Non-fibrous composites

Non-fibrous fillers include glass particles and a wide variety of mineral fillers. Often the loading in the polymer is high, making speciment preparation difficult as discussed in Chapter 2. Some novel methods have been tried, one of the more unusual being the production of thin sheets of material which are then clamped, heated, and expanded by blowing compressed air from below so as to produce a *bubble*. On cooling, the thin material in the bubble wall is selected for examination. Although bypassing the difficulty of sectioning, the technique is beset with other problems such as thickness control and local rapid thinning leading to a bursting of the bubble. Preparation by grinding and polishing surfaces remains the more usual method for these materials and is the only one available for thermosetting polymers. Such preparations are of course examined by reflected-light methods.

As well as common light techniques, crossed polars may be used to detect and identify mineral fillers and other doubly refracting components such as cellulose. Note that there may be alignment of the filler. This is common in the case of talc, for example, and means that the specimen needs to be rotatable on the microscope stage as a filler particle in an extinction position can be easily missed.

6

Molecular orientation in polymers

6.1. Introduction

The concept of molecular orientation in polymers has been outlined in Chapter 1 where it was intimated that image contrast could be obtained using a polarizing microscope. It is possible to go further and quantify the double refraction effect using techniques of birefringence measurement developed by petrologists and mineralogists. Molecular orientation on all but the smallest scale can have a profound influence on mechanical properties. This influence may be desirable, having been intentionally produced during the processing of the polymer, or undesirable, being the result of unsatisfactory processing.

Fibres and thin films are good examples of instances where stiffness and strength are favourably modified by producing molecular orientation during production. In contrast, orientation in injection-moulded or extruded plastics can give rise to weakness and to premature failure in service. It is very common for mouldings to have different degrees and directions of orientation through their thickness (note the similarity with the comments on glass-fibre orientation made in Chapter 5).

In the case of amorphous polymer, the correlation between optical (birefringence) measurements and molecular orientation is comparatively straightforward. When the material is semicrystalline, however, a more complicated problem exists. We have to consider the crystalline regions and amorphous regions in the material separately. Unfortunately optical methods do not normally separate the two, so that microscopic birefringence measurements give only an average figure for the material. Separation can be achieved if data from other techniques are available. Thus X-ray diffraction data will give information on crystallite orientation which can be subtracted from the birefringence result to allow assessment of molecular orientation in the amorphous regions.

In practice it is often permissible to assume that crystallite and amorphous orientations are similar and to use a birefringence measurement to characterize the plastics product. This is especially the case when dealing with films and fibres.

A further complication with certain composites and semicrystalline polymers is the role of form birefringence mentioned in Chapter 5. Again this is added in to a birefringence measurement. Thus, in general,

$$\Delta n_m = (1 - x_c)\Delta n_a + x_c \Delta n_c + \Delta n_f$$

where Δn_m, Δn_a, Δn_c, and Δn_f are the measured birefringence and the amorphous, crystalline, and form birefringence components, respectively. x_c is the fraction of crystalline material present.

Difficulties arise when the way in which the components of birefringence are to be added is considered. These problems are outside the scope of this text but fortunately the Δn_f component is usually of little significance except when dealing with certain block copolymers or terpolymers.

The overall picture is that, in practice, molecular-orientation assessment of both amorphous and semicrystalline polymers is permissible by measurement of birefringence, but an appreciation of exactly what is being measured is essential. No precisely defined relationship between molecular orientation and birefringence is generally possible as the former is itself capable of definition in a number of ways and by a distribution function rather than a single figure.

6.2. Microscopy equipment

A transmitted-light polarizing microscope with the following features is necessary for birefringence measurement work on polymers.

1. A high-intensity light source;
2. A graduated rotating stage;
3. A rotating graduated analyser (needed with Senarmont compensator only);
4. Strain-free condenser and objective lenses;
5. Body slots for the insertion of compensators or retardation plates;
6. Provision for the insertion of narrow band filters.

The objectives used rarely exceed × 20 and the numerical aperture is largely unimportant. The provision of a Bertrand lens or an alternative method of obtaining an enlarged view of the back focal plane of the objective, such as a telescope, can be of use when examining polymer films.

6.3. The aims of birefringence measurements

The refractive optical properties of a homogeneous material can be represented in the most general case by three principal refractive indices α, β, and γ which are mutually at right angles to one another. To optically characterize the material we need to know

1. The vibration directions (*principal* directions) of light waves passing through the material. The speed of these waves will be governed by the principal indices. The directions are referred to some *reference* direction such as the edge of the specimen.

2. The magnitude of α, β, and γ. In working with a polarizing microscope it is usual to measure the difference between the principal indices, i.e. either $(\gamma - \beta)$, $(\gamma - \alpha)$, or $(\beta - \alpha)$. These are the principal birefringes shown by the specimen and

they give us information on molecular orientation. Individual indices may be measured by the Becke Line Test using plane-polarized light (*not* crossed polars), or by microinterferometry. The Becke line method is limited to application only at the edge of features in the microscopic image and also requires the specimen to be reasonably homogeneous. Neither this method nor the interferometric technique are explored further here.

For an isotropic (unoriented) material, $\gamma = \beta = \alpha$, so that no birefringence is shown. Anisotropic (oriented) materials may require all three indices to describe them, if biaxial, or possibly only two, if there is greater symmetry in the molecular arrangement giving the uniaxial case.

Polymer fibres are 'drawn' or stretched in one direction only during manufacture and are usually uniaxial in optical character. Films may be uniaxial or biaxial according to their method of manufacture.

When molecular orientation in a fibre is produced by drawing it is found that as the percentage extension is increased so is the fibre's birefringence, but only up to a certain limit. Almost full alignment of the molecules has been obtained at this point and further extension can only be achieved by chain slippage. The maximum birefringence that can be shown by a polymer is a function of chemical composition and the arrangement of atoms in the chain. Fig. 6.1 shows the birefringence of a drawn nylon 6 fibre as a function of extension.

Comparing the magnitude of the birefringences of fibres of the same polymer allows us to compare the relative degrees of molecular orientation. Comparisons with the maximum possible birefringence Δn_{max} allow an absolute result to be obtained. Fibres are easier to consider in this way since they are represented by the simpler uniaxial case. However similar arguments may be used to understand molecular orientation in films and mouldings where the pattern of orientation is usually biaxial and more complex.

6.4. Compensators and compensation

The colours which may be seen when a piece of transparent polymer is examined between crossed polars are a manifestation of optical path differences in the material. At any given point in the specimen

$$\text{Optical Path Difference} = \text{OPD} = t\Delta n$$

when t is the geometrical thickness of the specimen and Δn the birefringence *displayed for that particular direction of observation*. This may or may not be a principal birefringence.

Reference to a Michel–Levy colour chart (often found in texts on polarized light microscopy) enables the colour to be converted into an OPD and then, if the thickness is known, into a birefringence. This procedure is subjective and therefore not fully satisfactory. To improve the accuracy of OPD measurement compensators are employed. It should be added however that it is good practice to

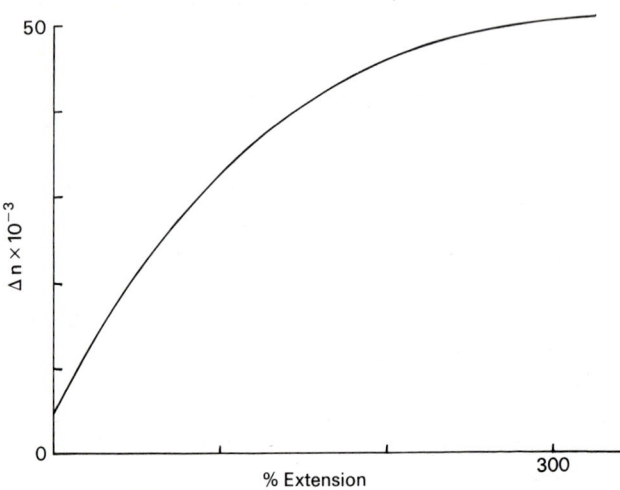

Fig. 6.1. Birefringence of nylon 6 fibre vs. extension.

always consult the Michel–levy chart before measurements are carried out in order roughly to predict the compensator result. Not only will this perhaps help with the selection of compensator but will also expose any slip in readings or calculation.

The optical path differences shown by polymer specimens can range from a few nanometres up to many micrometres. There are compensators available for dealing with any part of this range but none will conveniently cope with all of it. It is convenient here to divide compensators into *high-order* and *low-order* devices, the rather arbitrary dividing line being the ability to measure optical path differences of around $\frac{1}{2}\lambda$.

High-order compensators

High-order compensators interpose a variable optical path difference into the light beam passing through the microscope. By rotation of the specimen (or in some cases the slot by which the device is inserted into the microscope body) it can be arranged that the high and low refractive directions of the compensator are superimposed on the low and high index directions of the specimen. The result is a subtraction of optical path differences. When the optical path difference produced by the variable compensator is identical to that of the specimen, a zero path difference results. This *compensated* part of the image will usually appear black. Often the compensator design is such that this occurs only in a band across the field of view. This band is referred to as a *zero-order fringe*. On either side of this fringe will be a sequence of coloured fringes, assuming white light is being used.

If the orientation of the compensator is such that its higher refractive index is aligned with that of the specimen, no compensation can occur since the optical path differences are added. Should this occur during measurements, it is an indication that rotation of the specimen (or compensator) through 90° is necessary before measurements can be carried out.

Note that it is necessary for the specimen and the compensator to be used with well-crossed polars and with the former in its position of maximum brightness. This will occur midway between, or at 45° to, the positions in which the specimen extinguishes between crossed polars. In practice the specimen is rotated to extinction without the compensator inserted, a position which can be accurately set, then rotated a further 45°. On inserting and adjusting the compensator the zero-order fringe is sought. If this cannot be found, the specimen is counterrotated 90°. The optical path difference generated by the compensator is obtained by reading a scale or dial on the device. This reading can be converted into an OPD by calibration or reference to manufacturers tables.

To calculate the specimen birefringence it is necessary to know its thickness. Inaccuracies in this measurement can be the biggest source of error or doubt in a birefringence assessment of a polymeric specimen.

Typical commercially available compensators include those attributed to Babinet, Berek, or Ehringhaus. The latter is available in a form capable of compensating optical path differences up to 80 micrometres and is widely used on polymer films and fibres. A simple calibrated wedge, positioned in the primary image plane of the microscope by insertion into a special eyepiece can be used for path differences of up to about 1.5 μm.

Low-order compensators

Two types of low-order compensators predominate in manufacturers catalogues. The elliptic or Brace–Kohler type consists of a doubly refracting plate, having accurately known optical path difference, rotating in its own plane. The theory behind these devices relies on the phase angle (ϕ) between the two plane-polarized waves emerging from the specimen being so small that the approximation $\sin \phi = \phi$ is valid. This equates to a maximum OPD of about $\lambda/10$.

As before the specimen is turned to its 45° position before measurements are made by rotating the compensator plate until the region of the specimen of interest is at maximum darkness.

The de Sénarmont compensator is commercially popular because it consists only of a fixed $\lambda/4$ plate positioned in the instrument so that its higher index or *slow* direction is parallel with the vibration direction of light leaving the polarizer. Note that an ordinary $\lambda/4$ plate inserted in the standard 45° body slot of a microscope will not normally fulfil this condition. Either the slot position or the polarizer must be rotated 45° or, more commonly, a special plate is provided with the slow axis at 45° to the insertion direction.

The angular phase difference between waves emerging from the specimen is measured directly by rotation of the microscope's *analyser*. It is therefore essential that this is of the type carrying a graduated scale.

Although all compensators demand the use of near-monochromatic light for precise measurement, this is particularly the case for the Senarmont compensator, its $\lambda/4$ optical path difference obviously being accurate only for the wavelength for which the compensator was designed.

Strictly the Senarmont compensator is a *fractional-order compensator*. This implies that it will, for *any* specimen OPD, provide a numerical result corresponding to the difference between the actual OPD of the specimen and the nearest whole number (or integral number) of wavelengths OPD. Thus an identical result will be obtained for OPDs of x, $\lambda + x$, $2\lambda + x$ as well as $\lambda - x$, $2\lambda - x$ etc. Thus in practice, with caution and with the assistance of a Michel–Levy colour chart or other method of assessing the number of whole wavelengths OPD, the Senarmont compensator may be used to measure a large OPD; normally however this compensator would be used for a specimen showing an OPD of $\lambda/2$ or less.

6.5. The dispersion problem and possible solutions

The principal refractive indices of an anisotropic specimen vary with wavelength. More important from the practical standpoint is that the difference in refractive indices, i.e. the birefringence, shown by sections of the material is also a function of wavelength. This manifests itself in observations as a departure from the normal polarization colours shown by the Michel–Levy chart. For small optical path differences, the polarization colours assume hues, often browns and shades of green, which are definitely not shown by the colour chart.

For image-contrasting purposes these anomalous colours are as useful as normal ones, but problems arise when quantitative results are required. The zero-order 'black' fringe does not appear black, and furthermore several fringes may assume similar colours. The use of a filter to give monochromatic light may be seen as a solution. This is not the case for large optical path differences, since it becomes difficult to count the large number of black fringes which traverse the specimen and the true zero order cannot be identified. A further complication is that the dispersion characteristics of the compensator itself will almost certainly not match that of the specimen. This is again a problem of high-order compensators used for measuring large optical path differences.

Several methods have been suggested for dealing with this problem in relation to polymers. One such method involves preliminary experiments in which specimens are strained in small steps and simultaneously observed in the polarizing microscope (commercially available equipment can be used for this). Fringes can be counted and their colours noted. These data can then be used to construct a correction graph relating true optical path differences to those which would be arrived at using the most conspicuous dark fringe. Errors up to 20 per cent in OPD have been recorded.

A more direct method, and one to which most polymers readily lend themselves, is to cut a tapered edge to the specimen (films and fibres can usually be cut with a razor blade) and to carry out measurements just at the top of the wedge so produced.

Monochromatic light is used and the first fringe to appear at the bottom of the wedge is watched for as the compensator is adjusted. This is then moved up the wedge by further adjustment. The compensator is read when this fringe arrives at the wedge top. It is interesting to carry out the same experiment using white light; what starts as a black zero-order fringe at the base of the wedge is highly-coloured when it reaches the top.

6.6. Birefringence measurements on fibres

The majority of synthetic fibres are uniaxial in nature and can be characterized by two principal refractive indices, $n_\perp$ and $n_\parallel$ which are respectively perpendicular to, and parallel to the axis of the fibre. Furthermore, their cross-sections are often regular in profile, allowing easy measurement of their thickness t and thus the calculation of birefringence Δn from OPD measurements. Between crossed polars, such fibres will show extinction when the axis of the fibre is perpendicular to, or parallel to the vibration direction of light coming from the polarizer. Note also that a cross-section cut at right angles to the fibre's axis will not show birefringence since observation is then along the direction of the optic axis. Oblique sections will show birefringence but this will be less than the maximum birefringence displayed when the fibre is viewed 'side-on'.

Fibres which have $n_\perp > n_\parallel$ are considered to be *negative* in optical sign and, conversely fibres having $n_\parallel > n_\perp$ are recognized as *positive*. The majority of synthetic fibres are positive.

The magnitude of the birefringence may, as described earlier, be taken as a measure of the degree of molecular orientation imparted to the polymer during fibre manufacture. Birefringence measurement may therefore be used to check manufacturing conditions, particularly the drawing stage of production, or for identification purposes. Note however that undrawn fibre may exhibit birefringence on account of

1. The presence of crystalline material in crystallizing polymers (e.g. nylon);
2. The small amount of molecular orientation imparted by the melt or solvent spinning part of the process.

Table 6.1 shows $n_\perp$, $n_\parallel$, and the birefringence of some selected common synthetic fibres.

6.7. Birefringence measurements on films

The optical characteristics of polymer films such as those widely used for packaging, photographic, and industrial applications, are more complex than those of fibres. Films are normally biaxial in nature and characterized by three principal refractive indices α, β, γ as described earlier. In practice the microscopists first task is to establish the principal directions in the film with reference to a fixed direction such as the cut edge of the film or the *machine direction* (MD). Normally

Table 6.1. *Some selected refractive index and birefringence data for synethetic polymer fibres*

Fibre	$n_\perp$	$n_\parallel$	Birefringence
Poly(vinylidene chloride)	1.52	1.51	0.01 (−)
Polyacrylonitrile	1.525	1.520	0.005 (−)
Acrylic	1.512	1.510	0.002 (−)
Cellulose triacetate	1.468	1.469	0.001 (+)
Undrawn nylon	1.52	1.54	0.02 (+)
Viscose rayon	1.525	1.545	0.02 (+)
Poly(tetrafluoroethylene)	1.34	1.39	0.05 (+)
Drawn nylon	1.52	1.58	0.06 (+)
Poly(ethylene terephthalate)	1.54	1.74	0.20 (+)
Polyethylene	1.52	1.55	0.30 (+)
Poly(p-phenylene terephthalamide)	1.64	2.35	0.71 (+)

two of the three principal axes will be in the plane of the film, with the third perpendicular to it.

Thickness measurement of film is easily carried out with a dial gauge or, if greater accuracy is necessary, by an interference method. Assessment of the *in the plane* birefringence Δn_p provides information on the *balance* of the film. A balanced film will tend to show similar properties (including mechanical properties) regardless of the direction of testing. A film which is perfectly balanced will have $\Delta n_p = 0$ and becomes uniaxial in character since two of the three principal indices must then be equal. *Out of the plane* measurements are practically more difficult and must use tilting techniques or thin sectioning of the film at right angles to principal axes in its plane.

Assuming the minimum index α is perpendicular to the film plane, the birefringences shown in sections perpendicular to α and β will be $(\beta - \alpha)$ and $(\gamma - \alpha)$, respectively. A measure of the degree to which molecules are constrained to be in the plane of the film (the *degree of planar orientation*) can be obtained by averaging these two birefringences, i.e.

$$\tfrac{1}{2}((\beta - \alpha) + (\gamma - \alpha)) = \frac{\gamma + \beta}{2} - \alpha.$$

Some typical results logged for a biaxially drawn polypropylene film are

1. γ along machine direction;
2. $(\gamma - \beta)$ in plane of film = 0.5×10^{-3};
3. $\dfrac{(\gamma + \beta)}{2} - \alpha = 10 \times 10^{-3}$.

These imply a fairly balanced film but with significantly high planar orientation.

6.8. Birefringence measurements on mouldings

Here the situation is yet more complex, even after putting aside the complications of separating crystalline and amorphous orientation effects. In general the molecular orientation at any point in a moulding will be biaxial in character. Variation

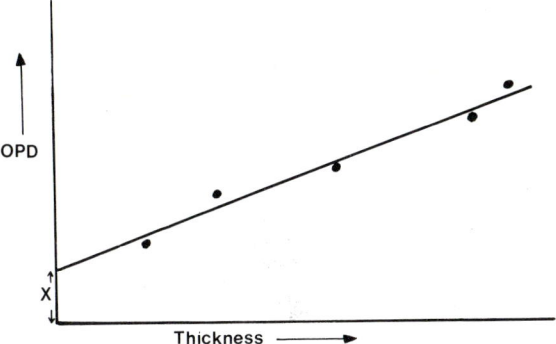

Fig. 6.2. OPD vs. thickness. The slope of the best straight line gives Δn. Intercept X on the OPD axis gives a measure of the cutting strain.

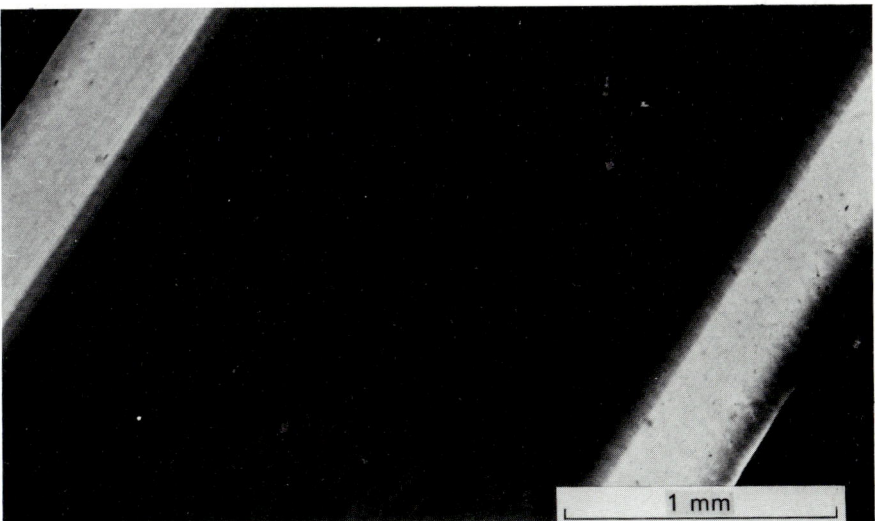

Fig. 6.3. Section of PVC moulding showing layers of differing molecular orientation.

in the directions of the principal axes and the magnitude of the birefringences displayed can be expected from place to place in the moulding. In particular variations may occur through the thickness. This can lead to a situation in which a full characterization of a moulding becomes a difficult and time-consuming business.

In practice,

1. Some regions of the moulding may be effectively uniaxial in character;
2. Only specific parts of the moulding (e.g. the surface layer) may be of importance;
3. Only a rough estimate of the degree of orientation may be required or only comparisons are needed.

These limitations substantially simplify the problem.

It is often tempting to characterize transparent mouldings by placing these between large crossed *Polaroid* sheets and dispensing with the microscope. Whilst this undoubtedly yields general information on molecular orientation in the moulding, it should be appreciated that there may be local variations through the thickness of the region being studied and that what is seen is some form of summation of the contribution of several layers.

Opaque mouldings, and transparent mouldings requiring more detailed study, are examined as thin sections. These are usually microtomed in two planes which are mutually perpendicular to allow full three-dimensional analysis. The extinction directions relative to a fixed direction such as a surface of the moulding are determined in the usual way, as is the OPD at points of interest. The section thickness is used to calculate birefringences from the OPD measurements.

Straining of sections during cutting can modify the birefringence displayed. Fig. 6.2 shows how this effect can be minimized. A series of different thickness sections are cut covering for example the range $2\,\mu$m to $20\,\mu$m. OPD is plotted as a function of thickness and the slope of the straight line obtained gives the birefringence. This procedure assumes that the cutting strain is consistent and independent of thickness. In practice this is found to be a reasonable assumption.

Fig. 6.3 shows a thin cross-section of a typical injection moulding showing different magnitudes of birefringence at different depths. The photomicrograph has been taken using circularly polarized light to eliminate extinction effects.

References

There are few books devoted to the light microscopy of synthetic polymers. The following list suggests books covering the basic principles of the technique. Selected chapters only will be of relevance in some of the more advanced texts.

Basic microscopy

Barer, R. (1953). *Lecture notes on the use of the microscope*. Blackwell, Oxford.
Bradbury, S. (1984). *Basic light microscopy*. RMS Microscopy Handbook Series, Oxford University Press, Oxford.
Hartley, G. (1979). *Hartley's microscopy*. Senecio Publishing Co., Oxford.
Martin, L.C. and Johnson, B.K. (1950). *Practical microscopy*. Blackie, London.
Moellring, F.K. (1972). *Beginning with the microscope*. Oak Tree Press, London.
Slayter, E. (1965). *Optical methods in biology*. John Wiley and Sons, New York.

Polarized-light microscopy

Gay, P. (1967). *An introduction to crystal optics*. Longmans.
Hallimond, A.F. (1970). *Manual of the polarising microscope*. Vickers Instruments, York.
Hartshorne, N. and Stuart, A. (1970). *Crystals and the polarising microscope*, 4th edn. Arnold, London.
Robinson, P.C. (1984). *Qualitative polarized light techniques*. RMS Microscopy Handbook Series, Oxford University Press, Oxford.
Wahlstrom, E.E. (1979). *Optical crystallography*, 5th edn. Wiley, New York.

Advanced microscopy

Francon, M. (1961). *Progress in microscopy*. Pergamon, Oxford.
McCrone Research Associated Inc. (1973). *Particle atlas*, Vols. 1–4. Ann Arbor Press, Michigan.
McLaughlin, R.B. (1975). *Accessories for the light microscope*. Microscope Publications, London.
Martin, L.C. (1973). *Theory of the microscope*. Blackie, London.
Rochow, T.G. and Rochow, E.G. (1978). *An introduction to microscopy by means of light, electrons, X-rays or ultrasound*. Plenum, New York.
Zieler, H.W. (1973). *The optical performance of the light microscope*. Microscope Publications, London.

Index

accessory plate, 41, 46
atactic polypropylene, 6

Becke Line Test, 66
Bertrand lens, 54, 68
birefringence, 5, 37; of fibres, 73, 74; of films, 73; form, 61, 62, 67; magnitude, 42; measurement, 67, 68; of mouldings, 74, 76; sign of spherulite, 41
blends, 52
block copolymers, 60

cast film structures, 35
chain folding, 6
chemical resistance, 18
circularly polarized light, 46
compensators, 66; high order, 70; low order, 71
composites, 56
cone and quartering, 12
contrast enhancement, 20, 36, 62; in composites, 58, 63; in phase contrast microscope, 47
copolymer, 3, 52
crystallinity, 5, 37, 73; degree of, 6, 8
crystallization structures, 28, 32, 33

differential interference contrast, of composites, 60; of spherulites, 48; of surfaces, 26, 30, 36
dispersion of birefringence, 72
domains, 3, 61
double refraction, 5, 67

extinction, 39, 45
extrusion defects, on extrudate, 28; on film, 33

fibre, birefringence, 70, 73; glass, 62
fibrils, 39, 40
fillers, 4, 66
film birefringence, 73
film surfaces, effect of additives, 35; optical character, 66; polyethylene, 33; polypropylene, 33
form birefringence, 61, 62, 67
fracture surfaces, 29

glass fibre, content, 64; length, 65; orientation, 63, 65; reinforcement, 22, 23, 62
glass transition, 7, 8

heterogeneous nucleation, 48
homogeneous nucleation, 48
homopolymer, 3
hotstage microscopy, design, 52; use, 50

isotactic polypropylene, 6
interferometry, of composites, 60; of surfaces, 32

knives, 15

lamellae, 6, 7; optical character, 38
light scattering pattern, 52

melting points, 52
melt pressing, method, 20; temperatures, 21

Michel–Levy colour chart, 66
microtome knives, 12
microtomy, 12
microvoids, 59
model of spherulite, 40
mounting, media, 20; specimens, 18

nucleation, additives, 49; row, 50; spherulite growth, 48

optical anisotropy, 38
optical melting point, 51, 52
orientation, glass fibre, 62; molecular, 67

phase contrast, 47
planar orientation, 74
polarizability, 4, 5
polarizing microscope, 27, 68
poly(ether sulphone), 25
polyethylene, 3, 13, 44; crystalline structures, 33, 37, 43; film surfaces, 33; light scattering, 52; moulding surfaces, 29; ringed spherulites, 45; sectioning, 16
polyformaldehyde, 42
poly(methyl methacrylate), 56, 57
poly(4-methyl pentene-1), 6, 38
polypropylene, 3, 6, 41, 42; film surfaces, 32, 33; glass filled, 22; mouldings, 41, 42; spherulite nucleation, 48, 49
polystyrene, 3, 18, 24; fracture surfaces, 30, 31
poly(tetrafluoroethylene), 25
poly(vinyl chloride)3, 56; sectioning, 16
powders, 15
processing of polymers, 8
propylene–ethylene copolymers, 13, 55

reflectivity of polymers, 25
refractive index, 3, 4, 9, 19; of blends, 56; variation with temperature, 8
ring morphology, 44

sampling, 11
section, quality, 13; thickness, 59
sectioning, 15, 16, 57
shadowing, 24
sheet moulding compound, 62
silica slides, 21
solvent casting, 20
specimen preparation, 10; brittle and hard polymers, 22; etchants, 22; physical properties of polymers, 11
spherulite, 6, 37; characterization, 39; extinction cross, 39; growth, 51; model, 40; polarization colours, 42; scattering pattern, 52; sign, 41, 42
stains, 20
stress, 45
styrene–butadiene–styrene copolymer, 61
surfaces, 24; extrudate, 26; film, 33, 35; fracture, 29; moulding, 26

tacticity, 6, 7
thermoplastics, 2
thermosets, 3, 66